INTERIM REPORT – TABLE OF CONTENTS

EXECUTIVE SUMMARY 4

PREAMBLE 8

SECTION I. THE BRAIN INITIATIVE: VISION AND PHILOSOPHY 10

    The Goal of the BRAIN Initiative 10

    Foundational Concepts: Neural Coding, Neural Circuit Dynamics

       and Neuromodulation 11

    Why Now? 12

    The Brain and Behavior 13

    Strategies and Experimental Systems 14

    Studying the Typical Brain Should Accelerate Understanding of Brain

       Disorders 15

    The Deliverables of the BRAIN Initiative 16

SECTION II. THE BRAIN INITIATIVE: SCIENTIFIC REVIEW AND
HIGH PRIORITY RESEARCH AREAS 17

    1. Mapping the Structure and Components of Circuits 17

    2. Neuronal Dynamics: Recording Neuronal Activity Across Time and

       Space 23

    3. Manipulating Circuit Activity 31

    4. The Importance of Behavior 32

    5. Theory, Modeling and Statistics Will Be Essential to Understanding

       the Brain 33

    6. Human Neuroscience and Neurotechnology 38

    7. Education 45

    8. Maximizing the Value of the BRAIN Initiative: Core Principles 47

CONCLUDING REMARKS 53

FURTHER READING 54

APPENDIX A 57

On April 2, 2013, President Obama launched the BRAIN Initiative to "accelerate the development and application of new technologies that will enable researchers to produce dynamic pictures of the brain that show how individual brain cells and complex neural circuits interact at the speed of thought." In response to this Grand Challenge, NIH convened a working group of the Advisory Committee to the Director, NIH, to develop a rigorous plan for achieving this scientific vision. To ensure a swift start, the NIH Director asked the group to deliver an interim report identifying high priority research areas that should be considered for the BRAIN Initiative NIH funding in Fiscal Year 2014. These areas of priority are reflected in this report and, ultimately, will be incorporated into the working group's broader scientific plan detailing a larger vision, timelines and milestones.

The goals voiced in the charge from the President and from the NIH Director are bold and ambitious. The working group agreed that in its initial stages, the best way to enable these goals is to accelerate technology development, as reflected in the name of the BRAIN Initiative: "Brain Research through Advancing Innovative Neurotechnologies." The focus is not on technology *per se*, but on the development and use of tools for acquiring fundamental insight about how the nervous system functions in health and disease. In addition, since this initiative is only one part of the NIH's substantial investment in basic and translational neuroscience, these technologies were evaluated for their potential to accelerate and advance other areas of neuroscience as well.

In analyzing these goals and the current state of neuroscience, the working group identified the analysis of circuits of interacting neurons as being particularly rich in opportunity, with potential for revolutionary advances. Truly understanding a circuit requires identifying and characterizing the component cells, defining their synaptic connections with one another, observing their dynamic patterns of activity *in vivo* during behavior, and perturbing these patterns to test their significance. It also requires an understanding of the algorithms that govern information processing within a circuit, and between interacting circuits in the brain as a whole. With these considerations in mind, the working group consulted extensively with the scientific community to evaluate challenges and opportunities in the field. Over the past four months, the working group met seven times and held workshops with invited experts to discuss technologies in chemistry and molecular biology; electrophysiology and optics; structural neurobiology; computation, theory, and data analysis; and human neuroscience (a full list of speakers and topics can be found in Appendix A). Workshop discussions addressed the value of appropriate experimental systems, animal and human models, and behavioral analysis. Each workshop included opportunity for public comments, which were valuable for considering the perspectives of patient advocacy groups, physicians, and members of the lay public.

Although we emphasize that this is an interim report, which will develop with much additional advice before June 2014, certain themes have already emerged that should become core principles for the NIH BRAIN Initiative.

1. *Use appropriate experimental system and models.* The goal is to understand the human brain, but many methods and ideas will be developed first in animal models. Experiments should take advantage of the unique strengths of diverse animal systems.
2. *Cross boundaries in interdisciplinary collaborations.* No single researcher or discovery will crack the brain's code. The most exciting approaches will bridge fields, linking experiment to theory, biology to engineering, tool development to experimental application, human neuroscience to non-human models, and more, in innovative ways.
3. *Integrate spatial and temporal scales.* A unified view of the brain will cross spatial and temporal levels, recognizing that the nervous system consists of interacting molecules, cells, and circuits across the entire body, and important functions can occur in milliseconds, minutes, or take a lifetime.
4. *Establish platforms for sharing data.* Public, integrated repositories for datasets and data analysis tools, with an emphasis on user accessibility and central maintenance, would have immense value.
5. *Validate and disseminate technology.* New methods should be critically tested through iterative interaction between tool-makers and experimentalists. After validation, mechanisms must be developed to make new tools available to all.
6. *Consider ethical implications of neuroscience research.* BRAIN Initiative research may raise important issues about neural enhancement, data privacy, and appropriate use of brain data in law, education and business. Involvement of the President's Bioethics Commission and neuroethics scholars will be invaluable in promoting serious and sustained consideration of these important issues. BRAIN Initiative research should hew to the highest ethical and legal standards for research with human subjects and with non-human animals under applicable federal and local laws.

The following research areas are identified as high-priority research areas in FY 2014.

#1. Generate a Census of Cell Types. It is within reach to characterize all cell types in the nervous system, and to develop tools to record, mark, and manipulate these precisely defined neurons *in vivo.* We envision an integrated, systematic census of neuronal and glial cell types, and new genetic and non-genetic tools to deliver genes, proteins, and chemicals to cells of interest. Priority should be given to methods that can be applied to many animal species and even to humans.

#2. Create Structural Maps of the Brain. It is increasingly possible to map connected neurons in local circuits and distributed brain systems, enabling an understanding of the relationship between neuronal structure and function. We envision improved technologies—faster, less expensive, scalable—for anatomic reconstruction of neural circuits at all scales, such as molecular markers for synapses, trans-synaptic tracers for identifying circuit inputs and outputs, and electron microscopy for detailed reconstruction. The effort would begin in animal models, but some mapping techniques may be applied to the human brain, providing for the first time cellular-level information complementary to the Human Connectome Project.

#3. Develop New Large-Scale Network Recording Capabilities. We should seize the challenge of recording dynamic neuronal activity from complete neural networks, over long periods, in all areas of the brain. There are promising opportunities both for improving existing technologies and for developing entirely new technologies for neuronal recording, including methods based on electrodes, optics, molecular genetics, and nanoscience, and encompassing different facets of brain activity, in animals and in some cases in humans.

#4. Develop A Suite of Tools for Circuit Manipulation. By directly activating and inhibiting populations of neurons, neuroscience is progressing from observation to causation, and much more is possible. To enable the immense potential of circuit manipulation, a new generation of tools for optogenetics, pharmacogenetics, and biochemical and electromagnetic modulation should be developed for use in animals and eventually in human patients. Emphasis should be placed on achieving modulation of circuits in patterns that mimic natural activity.

#5. Link Neuronal Activity to Behavior. The clever use of virtual reality, machine learning, and miniaturized recording devices has the potential to dramatically increase our understanding of how neuronal activity underlies cognition and behavior. This path can be enabled by developing technologies to quantify and interpret animal behavior, at high temporal and spatial resolution, reliably, objectively, over long periods of time, under a broad set of conditions, and in combination with concurrent measurement and manipulation of neuronal activity.

#6. Integrate Theory, Modeling, Statistics, and Computation with Experimentation. Rigorous theory, modeling and statistics are advancing our understanding of complex, nonlinear brain functions where human intuition fails. New kinds of data are accruing at increasing rates, mandating new methods of data analysis and interpretation. To enable progress in theory and data analysis, we must foster collaborations between experimentalists and scientists from statistics, physics, mathematics, engineering and computer science.

#7. Delineate Mechanisms Underlying Human Imaging Technologies. We must improve spatial resolution and/or temporal sampling of human brain imaging techniques, and develop a better understanding of cellular mechanisms underlying commonly measured human brain signals (fMRI, Diffusion Weighted MRI (DW MRI), EEG, MEG, PET)—for example, by linking fMRI signals to cellular-resolution population activity of neurons and glia contained within the imaged voxel, or by linking DW MRI connectivity information to axonal anatomy. Understanding these links will permit more effective use of clinical tools for manipulating circuit activity, such as deep brain stimulation and transcranial magnetic stimulation.

#8. Create Mechanisms to Enable Collection of Human Data. Humans who are undergoing diagnostic brain monitoring or receiving neurotechnology for clinical applications provide an extraordinary opportunity for scientific research. This setting enables research on human brain function, the mechanisms of human brain disorders, the effect of therapy, and the value of diagnostics. Meeting this opportunity requires closely integrated research teams including clinicians, engineers, and scientists, all performing according to the highest ethical standards of clinical care and research. New mechanisms are needed to maximize the collection of this priceless information and ensure that it benefits people with brain disorders.

#9. Disseminate Knowledge and Training. Progress would be dramatically accelerated by the rapid dissemination of skills across the community. To enable the broadest possible impact of newly developed methods, and their rigorous application, support should be provided for training—for example, summer courses and course modules in computational neuroscience, statistics, imaging, electrophysiology, and optogenetics—and for educating non-neuroscientists in neuroscience.

Although these FY 2014 research priorities are presented as nine individual recommendations, the overarching vision is to combine these approaches into a single, integrated science of cells, circuits, brain and behavior. For example, there is immense added value if recordings from neuronal populations are conducted in identified cell types whose anatomical connections are established in the same study. Such an experiment is currently an exceptional tour de force; with new technology, it could become routine. In another example, neuronal populations recorded during complex behavior might be immediately retested with circuit manipulation techniques to determine their causal role in generating the behavior. Theory and modeling could be woven into successive stages of ongoing experiments, enabling effective bridges to be built from single cells to connectivity maps, population dynamics, and behavior. Facilitating this vision of integrated, seamless inquiry across levels is the initial goal of the BRAIN Initiative, to be explored and refined before the final report in June 2014.

PREAMBLE. THE GOALS OF THE BRAIN INITIATIVE

We stand on the verge of a great journey into the unknown—the interior terrain of thinking, feeling, perceiving, learning, deciding, and acting to achieve our goals—that lives within the human brain. These capacities are the essence of our minds and the aspects of being human that matter most to us. Remarkably, these powerful yet exquisitely nuanced capacities emerge from electrical and chemical interactions among roughly 100 billion nerve cells and glial cells that compose our brains. All human brains share basic anatomical circuits and synaptic interactions, but the precise pattern of connections and interactions are highly variable from person to person—and therein lies the source of the remarkable variation we see in human behavior, from the breathtaking dance of a ballerina, to the elegant craftsmanship of a master carpenter, to the shrewd judgment of an expert trader. Our brains make us who we are, enabling us to perceive beauty, teach our children, remember loved ones, react against injustice, learn from history, and imagine a different future.

The human brain is simply astonishing—no less astonishing to those of us who have spent our careers studying its patterns and mysteries than to those new to thinking about the brain. President Obama, by creating the BRAIN Initiative, has provided an unprecedented opportunity to solve those mysteries. **The challenge is to map the circuits of the brain, measure the fluctuating patterns of electrical and chemical activity flowing within those circuits, and understand how their interplay creates our unique cognitive and behavioral capabilities.** We should pursue this goal simultaneously in humans and in simpler nervous systems in which we can learn important lessons far more quickly. But our ultimate goal is to understand our own brains.

Like the Apollo program, this challenging objective will require the development of an array of new technologies, drawing on scientists and engineers from a multitude of disciplines. These technologies will have to be integrated in an unprecedented manner to achieve the initiative's goals. We are at a unique moment in the history of neuroscience—a moment when technological innovation has created possibilities for discoveries that could, cumulatively, lead to a revolution in our understanding of the brain. The new technologies described in this report are already laying a foundation for exceptional progress, but more innovation is required. New molecular techniques to identify the specific connections between nerve cells that change when a new memory is formed, or a new social situation encountered, would create spectacular opportunities for deeper understanding. Similarly, new physics and engineering methods for noninvasive measurement and tuning of activity in fine-scale human brain circuits would create a revolution in the understanding and treatment of disease.

What will be gained by solving the mystery of the brain's circuits and their activity across time and space? Understanding the brain is a riveting intellectual challenge in and of itself. But in the longer term, new treatments for devastating brain diseases are likely to emerge from a deeper understanding of brain circuits. For example, treatment of Parkinson's disease has been greatly enhanced by circuit-level understanding of the brain's motor systems. Our front-line treatment for Parkinson's is the dopamine precursor drug, L-dopa, but its efficacy decreases over time while severe side effects increase. In response, teams of neurophysiologists, engineers, and physicians fused an understanding of the brain's motor circuits with technological advances to create deep brain stimulation (DBS), which can restore motor circuit function in many Parkinson's patients for up to several years. Current research into brain circuits for mood and emotion has the potential to advance psychiatry in similar ways.

We believe this to be a moment in the science of the brain where our knowledge base, our new technical capabilities, and our dedicated and coordinated efforts can generate great leaps forward in just a few years or decades. Like other great leaps in the history of science—the development of atomic and nuclear physics, the unraveling of the genetic code—this one will change human society forever. Through deepened knowledge of how our brains actually work, we will understand ourselves differently, treat disease more incisively, educate our children more effectively, practice law and governance with greater insight, and develop more understanding of others whose brains have been molded in circumstances different from our own. To achieve this vision, our nation must train and support a new generation of trans-disciplinary brain scientists and provide the resources needed to unleash their creative energies for the benefit of all.

On a personal note, the members of this committee are grateful to President Obama and the NIH for the opportunity to embark on our own journey of discovery over the past several months in preparing this report. We are indebted to the numerous colleagues who participated in our four summer workshops or shared their insights in one-on-one conversations, arguing with us and educating us in the process. We also value the perspectives offered to us by patient advocacy groups and members of the scientific community and lay public. This journey has already proved lively and enjoyable. We look forward to the next phase of discovery.

Respectfully, the members of the BRAIN Initiative Working Group

SECTION I. THE BRAIN INITIATIVE: VISION AND PHILOSOPHY

On April 2, 2013, the White House proposed a major national project to unlock the mysteries of the brain—the Brain Research through Advancing Innovative Neurotechnologies (BRAIN) Initiative. The President called on scientists to "get a dynamic picture of the brain in action and better understand how we think and how we learn and how we remember." In response to the President's call to action, the Director of the National Institutes of Health created this Working Group to "catalyze an interdisciplinary effort of unprecedented scope" to discover the patterns of neural activity and underlying circuit mechanisms that mediate mental and behavioral processes, including perception, memory, learning, planning, emotion, and complex thought:

> "By exploring these patterns of activity, both spatially and temporally, and utilizing simpler systems to learn how circuits function, we can generate a more comprehensive understanding of how the brain produces complex thoughts and behaviors. This knowledge will be an essential guide to progress in diagnosing, treating, and potentially curing the neurological diseases and disorders that devastate so many lives."
>
> — Charge to the NIH BRAIN Working Group, April 2013

This ambitious "American Project", articulated eloquently by President Obama in a White House announcement, can only be achieved through innovative, multidisciplinary investigation at all levels of nervous system function—behavioral, electrophysiological, anatomical, cellular and molecular. In parallel, advances in theory, computation, and analytics will be essential to understand and manage the large quantities of new data that will soon flow from neuroscience laboratories.

Over the past five months, we have reviewed the state of the field and identified key research opportunities. In this initial report we recommend specific goals to guide the BRAIN Initiative in Fiscal Year 2014. Our final report, to follow in June 2014 will define the vision more sharply, make longer term recommendations, and suggest benchmarks for evaluating progress toward the goals.

The Goal of the BRAIN Initiative

Our charge is to understand the circuits and patterns of neural activity that give rise to mental experience and behavior. To achieve this goal for any circuit requires an integrated view of its component cell types, their local and long-range synaptic connections, their electrical and chemical activity over time, and the functional consequences of that activity at the levels of circuits, the brain, and behavior. Combining these elements is at present immensely difficult even for one circuit, yet we must also weave together the many interlocking circuits in a single brain. As the President said in his White House press conference, this is indeed a "grand challenge for the 21st century."

As for any field and any era, progress toward these scientific goals is limited, to a large extent, by the experiments that are technically possible. But we are now within a period of rapid—perhaps revolutionary—technological innovation that could vastly accelerate progress toward an integrated understanding of neural circuits and activity. Thus for this interim report, our planning effort embraces a substantial technology emphasis, as reflected in the name of the working group: "Brain Research through Advancing Innovative Neurotechnologies." Our focus is not on technology *per se*, but on the development and use of tools for acquiring fundamental insight about how the nervous system functions in health and disease. We have considered how mature technologies can be applied to neuroscience in novel ways, how new technologies of obvious relevance can be rapidly developed and integrated into regular neuroscience practice, and what longer term investments should be made in 'blue sky' technologies with higher risk but potentially high payoff. As the BRAIN Initiative progresses, these technologies should increasingly be used to shed light on the healthy brain and on tragic human brain disorders.

Developing these novel technologies will require intense, iterative collaboration between neuroscientists and colleagues in the biological, physical, engineering, mathematical, and statistical and behavioral sciences. Essential partners should come from the private sector as well: corporate expertise in microelectronics, optics, wireless communication, and organization and mining of 'big data' sets can radically accelerate the BRAIN Initiative. Finally, clinicians will be essential partners to translate new tools and knowledge into diagnostics and therapies. The new technical and conceptual approaches to be created as part of the BRAIN Initiative will exert maximal impact if accompanied by specific plans for implementation, validation and dissemination to a larger community. Catalyzing the necessary collaborations and delivering reliable tools and resources to neuroscience laboratories should be major, overarching themes of the BRAIN Initiative.

Foundational Concepts: Neural Coding, Neural Circuit Dynamics and Neuromodulation

Neural coding and neural circuit dynamics are conceptual foundations upon which to base a mechanistic understanding of the brain. At the microscopic scale, the brain consists of vast networks of neurons that are wired together with synaptic connections to form neural circuits. In an active brain, each neuron can have electrical and chemical activity that is different from that of its neighbors; thus some neurons can play specialized roles in different tasks. Yet the activity of each neuron also depends on that of the others in the circuit, through the synaptic connections that define the circuit's architecture. Synaptic connections can change strength as a result of recent activity in the circuit, meaning that circuit architecture is constantly modified by experience. A thinking brain can therefore be viewed as an immensely complex pattern of activity distributed across multiple, ever-changing circuits.

Neural coding refers to how information about the environment, the individual's needs, motivational states, and previous experience are represented in the electrical and chemical activity of the neurons in the circuit. In a familiar example, the neural code for color vision begins with just three basic detectors in the eye—the cone photoreceptors. Circuits in our brains combine patterns of cone activation with other inputs to discriminate over a million different colors. More sophisticated, and poorly understood, neural codes enable us to recognize instantly the voice of a friend or the dramatic light of a Rembrandt painting. Elucidating the nature of complex neural codes and the logic that underlies them is one goal of the BRAIN Initiative.

As different neurons become silent or active in a thinking brain, the pattern of activity shifts in space and time across different circuits and brain regions. These shifting patterns define what is known as neural circuit dynamics. A key to understanding how the brain works is to determine how the neural dynamics across these vast networks process information relevant to behavior. For example, what is the form of neural dynamics in a circuit that makes a decision? What are the dynamically changing patterns of activity for speaking a sentence or imagining a future action? To probe the mechanics of the brain more deeply, we must learn how the biophysical properties of neurons and the architecture of circuits shape dynamic patterns of neural activity and how these patterns interact with incoming sensory information, memory, and outgoing motor commands. In the same way that the basic electrophysiological properties of single neurons are common across brain areas and species, it is likely that many fundamental forms of neural dynamics will generalize as well. One goal of the BRAIN Initiative is the identification and characterization of universal forms of neural circuit dynamics, likely represented by dynamical motifs such as attractors, sequence generation, oscillation, persistent activity, synchrony-based computation, and others yet to be discovered.

Accompanying this rapid flow of information that drives cognition, perception, and action are slower modulatory influences associated with arousal, emotion, motivation, physiological needs, and circadian states. In some cases, these slower influences are associated with specialized neuromodulatory chemicals like serotonin and neuropeptides, often produced deep in the brain or even in peripheral tissues, that can act locally or globally to change the flow of information across other brain circuits. In effect, neuromodulatory modifications of synaptic efficacy can 'rewire' a circuit to produce different dynamic patterns of activity at different points in time. The BRAIN Initiative should strive for a deeper understanding of these powerful but elusive regulators of mood and behavior.

Why Now?

This is a propitious moment for a sustained national effort to unlock the secrets of the brain. The reason lies in the technological and conceptual revolution that is underway in modern neuroscience. New molecular, genetic and cellular tools are generating exquisite insights into the remarkably diverse neuronal cell types that exist within our brains, the basic 'parts list' of our neural circuits. Novel anatomical techniques are providing remarkable new opportunities for tracing the interconnections between brain regions and individual neurons, revealing basic brain circuit maps in unprecedented detail. Innovative electrical and optical recording tools are allowing us to measure the intricate patterns of electrical activity that exist within those circuits across a broad array of behaviors ranging from decision-making to memory to sleep. Only a short time ago, we were restricted to studying the brain's electrical activity one nerve cell at a time; now we can record from hundreds of nerve cells, allowing us to analyze the cooperative activity of nerve cells as they operate in intact circuits; we look toward a future in which we can measure even richer patterns of brain activity, involving millions of nerve cells at any instant. Furthermore, newly invented genetic and chemically based techniques are giving us the power to modify activity in those circuits with great precision, creating extraordinary opportunities for deciphering the information-carrying codes in patterned electrical activity, and in the longer term, creating a foundation for novel therapeutic treatments for disease.

With these increasingly powerful techniques come new data sets of massive size and complexity. Reconstructing neural circuits and their dynamic activity in fine detail will require image analysis at a formidable scale as well as simultaneous activity measurements from thousands of neurons. The age of 'big data' for the brain is upon us. Thus, neuroscientists are seeking increasingly close collaborations with experts in computation, statistics and theory in order to mine and understand the secrets embedded in their data. These startling new technologies, many of which did not exist 10 years ago, force us to reconceive what it means to be an experimental neuroscientist today.

The challenge that now faces neuroscience lies in integrating these diverse experimental approaches and scaling them up to the level of circuits and systems. Previously, we could study the brain at very high resolution by examining individual genes, molecules, synapses, and neurons, or we could study large brain areas at low resolution with whole-brain imaging. Continued progress at both of these levels is essential, but our unique new opportunity is to study the critical intermediate level as well—the thousands and millions of neurons that make up a functional circuit. Remarkable new discoveries are possible at this intermediate level, for here we expect to observe the circuits, codes, dynamics, and information processing strategies that enable a collection of nerve cells to generate a complex, organized behavior.

The Brain and Behavior

The purpose of the brain is to generate adaptive behavior—predicting, interpreting, and responding to a complex world. As foreshadowed in the preceding section, some of the most riveting questions in neuroscience revolve around the relationship between neural circuit structure, neural dynamics, and complex behavior. Objectively measureable behavior is an indispensable anchor for the field of neuroscience—it defines the set of phenomena that we ultimately seek to explain. We benefit in this respect from the rich traditions of experimental psychology, psychophysics and neuroethology, but new innovation is needed in the analysis of behavior. Dobzhansky once said that "Nothing in biology makes sense except in the light of evolution," and it is no exaggeration to say that nothing in neuroscience makes sense except in the light of behavior. Thus a primary theme of the BRAIN Initiative should be to illuminate how the tens of billions of neurons in the central nervous system interact to produce behavior.

In advanced organisms our concept of 'behavior' must be extended to include sophisticated internal cognitive processes, in addition to externally observable actions. This point is dramatized by the story of Jean-Dominique Bauby, a French magazine editor who was left in a 'locked-in' condition by a brainstem stroke. Bauby was robbed of all voluntary movement except the ability to blink his left eye. Using the one behavior left to him, he wrote <u>The Diving Bell and the Butterfly</u>, an astounding memoir of the rich internal mental life that he continued to experience after his stroke:

> "My diving bell becomes less oppressive, and my mind takes flight like a butterfly. There is so much to do. You can wander off into space or in time, set out for Tierra del Fuego or for King Midas's court. You can visit the woman you love, slide down beside her and stroke her still-sleeping face. You can build castles in Spain, steal the Golden Fleece, discover Atlantis, realize your childhood dreams and adult ambitions."

Mental life can flourish within the nervous system, even if the behavioral link to the observable world is tenuous. Thus the BRAIN Initiative should focus on internal cognitive processes and mental states in addition to overt behavior. Accordingly, a preferred experimental emphasis should be on whole animals (typically behaving animals) with a secondary emphasis on reduced circuits that maintain important connections and integrative properties.

Measuring internal cognitive processes in animals is challenging, but rigorous methods have been developed to assess perception, memory, attention, decision-making, reward prediction, and many other examples. Although we must be constantly on guard against facile anthropomorphism, the continuity of brain structure and organization across species provides confidence that some cognitive processes analogous to ours are likely to exist in the brains of animals other than humans. Improving the behavioral analysis of these cognitive processes, both in experimental animals and in humans, should be a central goal of the BRAIN Initiative.

Strategies and Experimental Systems

Ultimately, the goal of the Brain Initiative is to understand how the human brain produces cognition and behavior, and specific recommendations of this report involve human neuroscience. However, human brains are complicated and difficult to access experimentally both for ethical and practical reasons. To reach our goal of understanding the human brain, it is therefore vital that we also investigate simpler animal brains as model systems—some with behaviors as comparable as possible to humans, but some with nervous systems that are more experimentally tractable. We cannot satisfy all requirements with a single animal model; a range of experimental systems, from simple to complex, will be needed to make progress. Fortunately, many basic principles of neural organization and function are conserved across animal species, so that progress in understanding simple systems can accelerate understanding of more complex systems.

In both animals and humans, we should enumerate and describe the brain's component parts— the different types of neurons and glia—and we should map their precise anatomical connections to obtain an accurate circuit diagram. We should measure the dynamic activity of the cells in a circuit under a variety of conditions and across a range of behaviors, and we should manipulate this activity to test causal hypotheses about how circuit activity influences behavior. Finally, we will require computationally powerful ways to analyze and understand the mechanisms by which dynamic patterns of activity in neural circuits give rise to behavior. These individual elements are clear; the challenge is how to accelerate, facilitate and combine them for maximum impact.

Studying the Typical Brain Should Accelerate Understanding of Brain Disorders

While the primary goal of the BRAIN Initiative is an understanding of normal brain function, we expect this work to provide an essential foundation for understanding neurological and psychiatric disorders. The burden of brain disorders is enormous. All of us are touched, directly or indirectly, by the ravages of degenerative diseases like Alzheimer's and Parkinson's, thought disorders like schizophrenia, mood and anxiety disorders like depression and post-traumatic stress disorder, and developmental disorders like autism spectrum disorders. Brain disorders limit personal independence and place enormous demands on family and society. The knowledge gained in the BRAIN initiative offers the possibility of reducing this burden.

There is reason to think that many disorders result in part from circuit dysfunction. Epilepsy is best understood as a circuit disease, where instability in neuronal communication leads to uncontrolled excitation and seizures. We currently stabilize patients by treating them chronically with potent drugs, but pinpointing abnormal circuits with new technologies could aid in the prediction of seizures and the development of more precise, localized treatments to stabilize activity. Like epilepsy, mood disorders and thought disorders are episodic, with an unstable waxing and waning of symptoms over days or years. The ability of many affected individuals to function normally at some times, and the absence of massive loss of brain cells, suggests that there may be no immovable obstacle to recovery of stable cognitive or emotional processing -- the circuitry for information flow exists, but it is not always regulated correctly. Unfortunately, there has not been a fundamentally new class of drugs for psychiatric disorders since the 1970s, largely because we understand neither how the circuits work nor how the drugs act on them. There are some clues, however. For example, the drugs we have for depression—most of which affect the neurotransmitters dopamine, serotonin, and norepinephrine—appear to act in part by modulating the flow of information between subcortical and cortical brain areas. A greater understanding of these circuits and their regulation could advance our ability to diagnose and treat thought and mood disorders.

Similarly, a better understanding of brain circuits has the potential to provide new paths to the treatment of neurological disorders. The neurodegenerative disorder Parkinson's disease is caused by the loss of dopaminergic neurons; like many neurodegenerative disorders, it manifests itself at the level of single cells. Nonetheless, those dopaminergic neurons are circuit elements. Changing the flow of information through the damaged circuits of Parkinson's disease patients with deep brain stimulation can dramatically improve their motor symptoms, even after most dopaminergic neurons are lost. We anticipate that refined knowledge of motor circuits will make deep brain stimulation more efficacious, and enable its use for other motor disorders. For other neurodegenerative disorders like Alzheimer's disease and motor neuron diseases, it may also be possible to deliver a therapeutic benefit by mimicking the circuit effect of a permanently lost population of cells. But first those circuit effects must be discovered, and interventional tools for delivering the appropriate circuit-level effect must be designed and built.

We also envision new ways to repair physical damage to the brain. Stroke, traumatic brain injury and spinal cord injury result in the loss of sight or memory, paralysis, or the inability to communicate. By tapping into existing brain circuits with new stimulators and sensors, it may be possible to re-establish damaged brain pathways, or allow control of prosthetic limbs with brain signals. Such implantable devices may sound like science fiction, but they are already in development in a few patients. Their success is limited by our fragmentary understanding of the brain's codes and instructions; there is great potential for human benefit from knowing more about the brain.

The Deliverables of the BRAIN Initiative

The BRAIN Initiative will deliver transformative scientific tools and methods that should accelerate all of basic neuroscience, translational neuroscience, and direct disease studies, as well as biology beyond neuroscience. It will deliver a foundation of knowledge about the function of the normal brain, its cellular components, the wiring of its circuits, its patterns of electrical activity at local and global scales, the causes and effects of those activity patterns, and the expression of brain activity in behavior. Through the interaction of experiment and theory, the BRAIN Initiative should elucidate the computational logic as well as the specific mechanisms of brain function at different spatial and temporal scales, defining the connections between molecules, neurons, circuits, activity, and behavior.

This new knowledge of the normal brain should form a foundation for more advanced translational research into brain disease mechanisms, diagnoses, and therapies. It should serve our colleagues in medicine, biotechnology, engineering, and the pharmaceutical and medical device industries, providing fundamental knowledge needed to ameliorate the vast human burden of brain disorders.

In addition to accelerating biomedical knowledge and treatment of disease, the BRAIN Initiative is likely to have practical economic benefits in the areas of artificial intelligence and 'smart' machines. Our brains can rapidly solve problems in vision, speech and motor coordination that the most powerful supercomputers cannot approach. As we learn more about the principles employed by the brain to solve these problems, new computing devices may be devised based on the cognitive architectures found in brains. Information companies are already investing in brain-inspired algorithms to enhance speech recognition, text search and language translation; the economic value of neurotech industries could someday rival that of biotech.

Finally, we hope through the BRAIN Initiative to also create a culture of neuroscience research that emphasizes worldwide collaboration, open sharing of results and tools, mutual education across disciplines, and the added value that comes from having many minds address the same questions from different angles.

SECTION II. THE BRAIN INITIATIVE: SCIENTIFIC REVIEW AND HIGH PRIORITY RESEARCH AREAS

1. Mapping the Structure and Components of Circuits

What classes of neurons and glia are involved in a given mental process or neural activity state? Which cells and brain regions contribute to a single percept or action, and how are they connected to each other? To answer these questions, we must define the cellular components of circuits, including their molecular properties and anatomical connections. This knowledge will tell us what the brain is made of at molecular, cellular, and structural levels; it will also provide a foundation for understanding how these properties change across the normal lifespan and in brain disorders.

1a. Cell Type: A Starting Point for Intellectual and Technical Progress

1a-i. A Census of Neuronal and Glial Cell Types

The brain contains many classes of neurons and glia, but not infinitely many. It consists of neurons that are distinguished by their neurotransmitters, electrophysiological properties, morphology, connectivity, patterns of gene expression, and probably other functional properties. These properties are major determinants of system-wide neural activity patterns. Classification of neurons is prerequisite to manipulating them in controlled ways, and to understanding how they change in brain disorders. Information about the types of glial cells, vascular cells, and immune cells associated with the nervous system may also increase our understanding of normal brain function and disease. Therefore, a valuable short-term goal for a BRAIN project is to generate a census of cell types within the brain.

There is not yet a consensus on what a neuronal type is, since a variety of factors including experience, connectivity and neuromodulators can diversify the molecular, electrical and structural properties of initially similar neurons. In some cases, there may not even be sharp boundaries separating subtypes from each other. Nonetheless, there is general agreement that types can be defined provisionally by invariant and generally intrinsic properties, and that this classification can provide a good starting point for a census. Thus, the census should begin with well-described large classes of neurons (e.g. excitatory pyramidal neurons of the cortex) and then proceed to finer categories within these classifications. This census would be taken with the knowledge that it will initially be incomplete, and will improve over iterations. A census of cell types is an important short-term goal for several reasons:

a) An agreed-upon set of cells provides a frame of reference for studies in many labs, and possibly in different organisms, allowing cross-comparisons. For example, to the extent that neuronal types are conserved across species (itself an important question) we can ask whether there are differences in their numbers and ratios in the cortex of primates compared to rodents.

b) An agreed-upon set of cells provides a foundation for further experiments, and shapes the problem going forward. For example, what genes are expressed in each of the different cell types? Do the sets of cells add up to 100%, suggesting that all neurons and glia are accounted for? Are there genetic elements such as Cre lines or viruses that provide experimental access to each cell type?

c) This problem can be solved with readily achievable improvements to existing technology.

d) This project has an initial endpoint, and the list itself will be a resource that can serve to organize subsequent BRAIN experiments and data analysis in a systematic way.

We envision a "neuron-ontology" bioinformatic framework that might be analogous to the gene-ontology framework in molecular genetics (http://www.geneontology.org/). The framework should include the potential to cross-reference information about homologous cell types from different animals, which could be valuable in the same way that information about homologous genes is valuable for linking studies across species.

The atlas should expand to describe the detailed morphology and connectivity of each neuronal class, its activity under different conditions, and its response to perturbations, as these results emerge. It could grow to include information about cells from human patients and animal models of human disease, extending its reach and providing insight into pathogenic processes.

The ultimate census of cell types would include all of the neurons and glia with molecular annotation at subcellular resolution: not just mRNA expression but ion channels, synaptic proteins, intracellular signaling pathways, and so on. This is beyond the reach of current technology, but stating the goal will provide impetus to technological development. The Allen Brain Atlas for RNA *in situ* hybridization is an example of the annotation of an anatomical database with molecular information. Similarly, array tomography can provide information about the location of specific proteins within cells based on antibody staining and optical imaging.

A census and database of neuronal cell types might begin with the mouse, where many genetic tools have already been developed and substantial data exist on gene expression patterns. Over the longer term, the census could be extended to different animal species and to humans.

1a-ii. Tools for Experimental Access to Defined Cell Types

The ability to define, monitor and manipulate a circuit requires experimental access to the individual cells and groups of cells within that circuit. Development of such tools will be facilitated by molecular analysis of cell types (section 1a-i), and should in turn facilitate progress in mapping neuronal connectivity (section 1b), understanding neuronal dynamics (section 2), and establishing function through causal neuronal manipulation (section 3).

The past decade has seen the development of remarkable genetic tools including calcium indicators (e.g. GCaMP), optogenetic tools (e.g. Channelrhodopsin), synaptic monitors (e.g. SynaptopHluorin), pharmacogenetic tools (e.g. RASSLs/DREADDs), and a variety of tags that permit proteins to be visualized *in vivo*. By their nature, using these tools requires the ability to deliver a gene to a neuron or neurons of interest ('genetic access'). Projects such as the HHMI *Drosophila* project, the NINDS GenSat project, and the Allen Brain Institute project have been, or are currently engaged in, developing genetic access to defined cell types in *Drosophila* and the mouse, but while these tools are nearing completion in the fly, they are not comprehensive in any other species.

The current methods for neuron-specific gene delivery in the mouse are typically bipartite, with (1) a recombinase gene such as Cre and (2) an effector/sensor gene that is activated by the recombinase. These two elements are independently delivered as transgenes, through recombination into an endogenous locus or insertion into bacterial artificial chromosomes (BACs), or as stereotactically-injected viruses; the "intersection" between the two elements generates a more restricted expression pattern than either one alone. The potential value of such lines is high, but there is much room for improvement, and only a small number are in regular use in the current literature.

In addition, non-genetic methods could be used to deliver active agents to neurons of particular types, and would expand the range of possible experiments. Viruses or liposomes that contain pharmacological agents, proteins, or nanoparticles might be coated with antibodies that direct them to certain cell types. Providing reliable access to specific cell types in particular neural circuits or brain areas will accelerate all areas of modern neuroscience.

A full exploration of methods for targeting genes, proteins, and chemicals to specific cell types is highly desirable. One question that should be considered during the initial stages of the BRAIN Initiative is whether genome engineering by conventional transgenesis could be superseded by methods that are faster, cheaper, and more easily generalized across species. Mouse husbandry is slow and expensive, and generating the right multi-transgenic strains is a financial and temporal drag on the progress of neuroscience research. Furthermore, we wish to study other species as well. The BRAIN Initiative should solicit new ideas for cell-type specific delivery of transgenes, perhaps based on viruses bearing small specific regulatory regions for intersectional cell-type definition via (for example) multiple recombinases; or viruses driving efficient, specific integration of exogenous genes into the genome through cutting-edge tools such as CRISPRs and TALENs; or antibody-targeted liposomes. Completely new ideas might emerge to address this problem. The most valuable ideas will be those with potential to solve the general problem for any species, in preference over those that would work for only one species at a time. The long term vision is development of comprehensive, general suites of tools that target expression to a brain area of interest, disseminated for broad, effective use in neuroscience labs around the world.

The next frontier would be gaining access to the human brain, which is more likely to involve transient delivery of RNA or a chemical than permanent genetic change, although viral vectors for human gene therapy are currently under exploration in the brain. Several pharmaceutical companies are developing tagged antibodies that cross the blood-brain barrier (e.g. via transferrin receptors), and these might be chemically or genetically engineered to include effectors or sensors of neuronal activity.

High Priority Research Area for FY2014: Generate a Census of Cell Types. It is within reach to characterize all cell types in the nervous system, and to develop tools to record, mark, and manipulate these precisely defined neurons *in vivo*. We envision an integrated, systematic census of neuronal and glial cell types, and new genetic and non-genetic tools to deliver genes, proteins, and chemicals to cells of interest. Priority should be given to methods that can be applied to many animal species and even to humans.

1b. The Structural Map: Tracing Anatomical Circuits at Different Scales

Rapid information flow across the brain is mediated by anatomical connections between cells, including local connections within a brain region and long-range connections into and out of that region. Defining circuit function requires knowledge of circuit structure. Three levels of anatomy should be considered: long-range, intermediate-range, and detailed connectivity.

1b-i. Long-Range Connectivity

Traditional neuroanatomy has focused on large-scale, long-range connections between different brain regions (e.g. the thalamocortical tract). In humans, long-range connections are being studied within the Human Connectome Project by noninvasive imaging methods. In animals, long-range connections are being pursued in detail using serial sectioning combined with modern dye-tracing techniques and genetic markers, with newly emerging whole-mount imaging and staining methods such as CLARITY and clearing techniques such as Sca*l*e and SeeDB poised to make an impact. Currently, most effort is being expended on the mouse model system, but these techniques can and should be extended to other species as well. The new whole-mount methods also appear promising for tracing tracts in human post-mortem tissue, and may provide an important high-resolution complement to noninvasive imaging methods.

"Projectomes" of this kind are attainable within the next few years with current and emerging technology. The next steps are identifying gaps and completing studies of rodent, non-human primate, and human anatomical tracts at high quality, integrating the results across labs and institutions, and making the information broadly available to the community. Inclusion of other species for comparative purposes is highly desirable. Combining these datasets into a common bioinformatic framework, and registering these datasets with other streams of information describing the cell populations and projections of interest, such as molecular phenotype and activity patterns during behavior, will increase their depth and scientific utility.

1b-ii. Intermediate-Scale Connectivity

The next problem is mapping circuits at an intermediate scale. What long- and short-range projections make up a specific functional circuit, which may consist of only some of the cells in a particular brain region? For example, brain regions such as the hypothalamus consist of mixed cell populations, each of which has very different input and output connections that are not evident in the large-scale connectivity. Mapping these connections currently requires considerable time and effort. Progress in this area is attainable and should be vigorously pursued.

There is considerable potential for improving the tools for studying intermediate-level circuitry. Trans-synaptic tracing of connections is highly desirable, but existing methods (lectins, dyes, and rabies-based viral tracers) are imperfect. For example, rabies tracers are largely limited to retrograde tracing, even though anterograde tracing is equally important for defining circuits. There is a concern that the present tracers may work on only a subset of cell types, and there is no fully accepted answer as to whether these tracers are strictly trans-synaptic or more generally trans-neuronal.

Better methods for tracing circuits are critical to rapid progress throughout neuroscience, and would provide important structural constraints for interpreting virtually all functional studies. Better trans-synaptic tracers would be extremely valuable, and their development should be encouraged by the BRAIN Initiative. These may be based on viruses or on different kinds of transgenic technologies; combining these methods with tract tracing or array tomography would increase their resolution. Other potential techniques are being explored, but none has yet matured: methods that use fluorescent proteins to label synapses (e.g. GRASP); or enzymatically-based detectors of trans-synaptic recognition (e.g. ID-PRIME), which have the enormous advantage of amplifying signals to enable robust detection. Truly transformative technologies could be encouraged from molecular biology or chemistry. The most attractive methods would be those applicable to many species, including humans. Methods that work in post-mortem brains would be particularly valuable for high resolution mapping of human brain circuits. Identifying intermediate-scale circuits should be a significant goal of the BRAIN Initiative.

1b-iii. Detailed Connectivity: Towards a Full Connectome

Finally, there is the question of reconstruction of circuits at very high resolution through electron microscopy, which is widely considered to be the gold standard for circuit mapping. To date sparse reconstruction has been used to examine small numbers of neurons in a variety of systems, but dense reconstruction has been applied only to the very small animal *C. elegans,* or to small parts of the nervous systems of larger animals such as *Drosophila;* ongoing studies in mammals are extending this approach to the retina. The past few years have seen great strides in sectioning and image collection techniques, but even so, electron microscopy is prohibitively slow for large-scale studies. The bottleneck is data analysis, the painstaking and potentially error-prone process of tracing fibers and mapping synapses from one very thin section of a brain to the next across thousands of successive sections.

The impact of dense EM reconstruction would be amplified tremendously if it were possible to increase throughput 100- or 1000-fold across all steps of the procedure, including segmentation and reconstruction as well as sectioning and data acquisition. Some promising improvements have been demonstrated, including automated capturing of serial sections for transmission electron microscopy, and serial block face scanning electron microscopy that maintains perfect

3-D registration during automated sectioning and data acquisition, but much remains to be done.

Possible areas for incremental improvement include:
1) Improved methods for the histological preparation of neural tissue, especially large samples. Can we engage chemists and chemical engineers in the problem to bring fresh approaches to this century-old area of research?
2) Improved methods for automated tissue sectioning and imaging, although this area has already progressed greatly.

Areas where progress is most needed are:
3) Improved software methods for segmenting and assembling the data. Technological advances in machine learning, artificial intelligence, and crowd-sourcing approaches to reconstruction could have a profound impact on the field.
4) Improved methods for synapse identification, in particular the ability to assess the type of synapse (excitatory, inhibitory, modulatory, electrical) and estimate synaptic strength. Cell type-specific markers or molecular markers of subsets of synapses that are visible at the EM level could be very helpful in large-scale reconstructions.

Electron microscopy is labor intensive, but it happens in stages. Sectioning for EM is relatively quick; scanning takes ten times as long; reconstructing is slower by orders of magnitude. If high quality scanned images were made available on the internet, individual users could spend their own time reconstructing areas of the brain of relevance to them, using software tools made available by the experts. Under this model, the laborious reconstruction task would be performed as-needed by a world-wide community of collaborators. The scope and impact of electron microscopy could be broadened beyond the relatively small group of expert labs by encouraging sharing of primary scanned images and reconstruction tools. There is no reason, in the modern era, for EM micrographs to be trapped in the lab that generated them. It would be exciting for the BRAIN Initiative to generate the basic data resource (high quality micrographs) for a variety of brains and species, with entirely open access to the data.

Truly innovative approaches to dense reconstruction should be encouraged, with a focus on the data analysis bottleneck and greatly improved throughput. A 100- or 1000-fold improvement should be held up as a serious goal. As with other approaches to wiring, registering these dense connectivity datasets with molecular phenotypes and activity patterns during behavior will vastly increase the scientific utility and interpretability of the data. A decision about whether, when, and how to scale up these anatomical approaches could be made after a comparison of new methods.

Development of these new technologies should proceed hand-in-hand with application to important problems in neuroscience. In the best case, dense reconstruction could be performed after recordings of neuronal activity and behavior in the same animal. The larval zebrafish is a promising system for a full, dense reconstruction of a vertebrate nervous system. Smaller projects in the mammalian retina, hippocampus, or cortex could have a broad impact. The important point is that broad support for large-scale, dense connectomics will only appear when it begins to yield answers to specific scientific questions that could not have emerged by other means.

High Priority Research Area for FY2014: Create Structural Maps of the Brain. It is increasingly possible to map connected neurons in local circuits and distributed brain systems, enabling an understanding of the relationship between neuronal structure and function. We envision improved technologies—faster, less expensive, scalable—for anatomic reconstruction of neural circuits at all scales, such as molecular markers for synapses, trans-synaptic tracers for identifying circuit inputs and outputs, and electron microscopy for detailed reconstruction. The effort would begin in animal models, but some mapping techniques may be applied to the human brain, providing for the first time cellular-level information complementary to the Human Connectome Project.

2. Neuronal Dynamics: Recording Neuronal Activity Across Time and Space

Understanding the electrical and chemical activity of neuronal circuits and systems is central to the BRAIN Initiative. The challenge is that these circuits incorporate neuronal activity at a variety of spatial and temporal scales. At the spatial level, an ensemble of neurons associated with a given behavioral task may be concentrated in one brain region, but not all neurons in that region may be part of the ensemble, and other important neurons will reside in different regions. For example, a circuit for conditioned fear behavior might include subsets of neurons in the primary sensory cortex and thalamus (for sensing the threat), the hippocampus (memory formation), the amygdala (fear learning), the autonomic nervous system (physiological output), and the prefrontal cortex (top-down control of behavioral response to the threat), among many others. To systematically study brain mechanisms underlying a particular behavior or cognitive process, it is important to sample neuronal activity broadly across brain structures and record from many identified cell types. It is also critical to measure and analyze neuronal activity at multiple time scales that are relevant to behavior and cognition: fast (e.g. spikes), intermediate (e.g. short-term plasticity, recurrent excitation) and slow (e.g. global attentional and arousal states; neuromodulation).

In an ideal world, a neuroscientist might or might not want to know the activity of every neuron in an animal under a given condition—this is a subject of debate—but there is general agreement that we need to measure neuronal activity across much larger spatial and temporal scales than we are managing at the moment. In the vast majority of experiments, we observe only a tiny fraction of the activity in any neuronal circuit, and then under a very limited range of behavioral conditions. How can we best identify the spatial and temporal patterns of activity that underlie specific cognitive processes and behaviors? How will we know when we have recorded from enough neurons to understand a cognitive process or mental state? What methods are needed to record all relevant kinds of activity in all relevant brain regions?

2a. What Neurons Should We Record? Identifying Dispersed Circuits

As illustrated above, it is important to scan broadly across the brain to identify distributed circuits. In general, distributed circuits must be defined functionally, based on activity of the constituent neurons during a behavior or under specific experimental conditions. Even the highest level of anatomical resolution is not sufficient to define a circuit, because synapses vary in their strength and modulation. An additional complication is that individual neurons may participate in different functional circuits under different experimental conditions or behavioral tasks. Mapping dispersed and overlapping circuits can be aided by labeling neurons that are active during a specific window of time, which permits identification of functionally related cells that are spatially intermixed with other cells.

Existing tools for identifying functional circuits on cellular scales must be improved, and development of novel tools strongly encouraged. The relationships between cells in circuits can be rigorously established by electrophysiological recordings in which one cell is stimulated and the other recorded, sometimes even with many neurons (e.g. laser scanning photostimulation), but this is not easily accomplished over long distances. A variety of methods involving optogenetic tools may help in this effort. For example, virally-delivered opsins and fluorescent proteins will spread throughout cells and down axons, allowing anatomically-defined optogenetic control by light delivery at the axon terminal region. Exciting defined cells or axon terminals with light while recording from a single postsynaptic cell can define sources of functional input (channelrhodopsin-assisted circuit mapping). This method in its current form is not equally effective in all settings, however, and is not easily combined with large-scale recordings.

One class of tools for circuit mapping is based on the expression of "immediate early genes" whose expression is up-regulated by sharp increases in neuronal activity. In its original form, each animal is only examined once, and the cells are dead and fixed by the time they are identified. In its modern form, immediate early gene expression can be coupled to reporters such as tamoxifen-regulated Cre recombinase, allowing permanent Cre-marking of neurons that were highly active at the time that tamoxifen was delivered or removed. While useful, the existing promoters are slow reporters with unpredictable regional and cell-type specificity, and their expression is only partially correlated with neuronal activity. In short, this is a tool to begin sketching a circuit for an entirely novel stimulus, but is not sufficient to watch ongoing, more modest changes in activity.

Improvements in methods to identify neurons in active circuits should be encouraged. The importance of these methods will be greatest for circuits that are distributed, or intermixed with other circuits, in a way that frustrates conventional anatomical tracing. Better time resolution is highly desirable. A new method uses phosphorylation of a ribosomal protein, S6, to label active cells while allowing their mRNA expression to be characterized; this method is applicable across mammals, not just to mice. Improved transcriptional reporters have been suggested that would require the coincidence of light activation and calcium entry to induce transcription, with light used to define the point at which activity is measured. There may be entirely new ways to solve this problem through biochemical or chemical reagents that mark active neurons. The ability to mark several circuits in series in the same animal would allow more sophisticated analysis, for example for within-animal comparisons of the effect of different behavioral states on neuronal responses.

2b. How Many Neurons Should We Record? The Test Case is Complete Circuits

Small systems provide a test bed for asking how much "emergent information" arises from recording an entire brain or brain structure, and provide initial clues to the density of recordings needed to characterize functional circuits. In a first example from the 1980s, voltage-sensitive dyes revealed widespread activation of over 100 abdominal ganglion neurons in the mollusc *Aplysia* during gill withdrawal behavior. However, other experiments argued that experience-dependent changes in gill withdrawal and its regulation could be controlled by just a few of these neurons. These contrasting observations posed important questions about the relative roles of population activity and single-neuron function that are still not answered. To assess the value added by monitoring all neurons in a circuit, complete or near-complete recordings should be gathered from a few model systems under a variety of conditions.

It is worth noting how far we still are from the goal of complete circuit recording, except for a few invertebrate ganglia, even though recording methods have been scaled up in recent years. For example, large-scale neurophysiological approaches have allowed recordings from thousands of neurons in the vertebrate retina, but even so only the retinal ganglion cells have been recorded at scale, not the many nonspiking bipolar and amacrine cells that process information prior to optic nerve output. Amacrine cells illustrate an additional challenge to the concept of "complete" recordings; some are thought to have subcellularly compartmentalized voltage signals that would be overlooked if recordings were made only from the soma.

A few test cases for large-scale, "complete" neuronal recordings would be of great interest, especially if gathered in close partnership with theory and behavioral analysis to provide context and interpretation. Genetically-encoded calcium indicators are already being used to image a large fraction of neurons in brains of the larval zebrafish and the nematode worm *C. elegans*, although not yet at speed in behaving animals. Complete recordings from neurons within well-defined mammalian brain areas are appealing, but will require new approaches.

2c. How Should We Record? Advancing Recording Technology

There are two important classes of methods for recording neuronal activity. Classically, electrophysiology with electrodes has been the workhorse of neuroscience. Microelectrode and macroelectrode recordings will continue to be important due to their high temporal resolution, their applicability to structures throughout the brain, and their appropriateness for human studies. More recently, optical methods for recording activity have been greatly improved, providing substantial opportunity for further advances. Both are important areas for development.

2c-i. Electrode Arrays for Recording Voltage and Passing Current

Micro- and macro-electrodes are widely used tools for recording neural voltage signals and stimulating neural tissue artificially via passage of electrical current. Understanding cognitive and behavioral processes mediated by distributed neural circuits will be greatly accelerated by the development of next-generation multi-electrode arrays that can record single cell activity simultaneously from large populations of neurons at multiple sites in the brain. New electrode architectures and accompanying methods for extracting the measured signals are being pursued in many laboratories. Here we highlight general problems whose solution would accelerate progress on many fronts, from basic circuit research in small animals to human clinical application.

Penetrating electrode arrays: Much recent work has been directed at developing arrays of penetrating microelectrodes using integrated circuit technology, with substantial efforts underway to increase the number of electrode shanks and the number of recording contacts per shank. Next-generation electrode arrays should also have the capability of simultaneous stimulation and recording from individual electrodes on the array, since this capability can be critical for establishing functional relationships among recorded neurons. Advances should be sought to address four primary problems currently impeding progress. First is the physical design of electrode arrays with greatly increased numbers of shanks and contacts; different designs will be needed for the complementary goals of making simultaneous recordings from very large numbers of neurons at one site (e.g. a cortical column) versus more modest numbers of neurons at many dispersed sites. Second is a major divide between "passive" and "active" electrode design. One obstacle to true high-density recording (e.g. >1000 recording contacts) is physical management of the large number of wires that must be attached to the recording contacts. This problem can be reduced substantially by incorporating "active" circuitry into the array implants that filters, amplifies, multiplexes and telemeters the primary signals close to the recording source. Solutions to these problems will require substantial engineering innovation in miniaturizing active electronics and reducing the power needs of the circuitry. Third is the continued development of hybrid electrode arrays that combine electrical and optical recording and stimulation capabilities, increasing experimental power relative to either method alone. Finally, particularly for human use, electrode arrays must be made more compatible with the target tissues in their mechanical compliance, cross-sectional area, lifetime in the implanted tissues, and immune tolerance.

Surface electrodes for recording average neural activity from cortical loci are an established technology with notable recent improvements: Small surface electrodes (20 micron diameter) at very high spatial densities (20 micron separation) have been created on thin, flexible parylene substrates (4 microns thick) that "conform" well to local curvature of the cortical surface. Such conformal arrays coupled to smart electronics would be particularly useful for research and device development (e.g. prosthetics) in humans.

Further progress will come from supporting a diverse assortment of penetrating and surface electrode array designs, with emphasis placed on designs that, 1) are smarter, smaller and use less power, 2) integrate multiple capabilities (e.g. electrical and optical; recording and stimulation), and 3) can be implemented at multiple scales, from small animals to humans. In the longer term, new materials and designs may prolong the useful lifetime of implanted arrays. Next-generation electrode array design is an area that might benefit substantially from interaction with private companies with expertise in IC chip design, miniaturization, wireless telemetry and low-power applications. Some companies with relevant expertise have expressed interest in partnering with the BRAIN Initiative. The main obstacle to such partnerships to date has been the lack of financial incentive due to the small market.

2c-ii. Optical Sensors of Neuronal Activity

The ability to monitor activity in large numbers of neurons has been accelerated over the past two decades by using optical methods and tools from chemistry and genetics. Optical sensors, whether chemical or genetic, have the potential to report sub-cellular dynamics in dendrites, spines, or axons; to probe non-electrical facets of neural dynamics such as the neurochemical and biochemical aspects of cells' activities; and to sample cells densely within local microcircuits. The capacity for dense sampling holds particular promise for revealing collective activity modes in local microcircuits that might be missed with sparser recording methods.

Genetic tools can also target cells by genetic type or connectivity, and maintain large-scale chronic recordings of identified cells or even individual synapses over weeks and months in live animals. Such large-scale chronic recordings are especially beneficial for long-term studies of learning and memory, circuit plasticity, development, animal models of brain disease and disorders, and the sustained effects of candidate therapeutics.

Although the acceleration in optical sensor development is relatively recent, it has already had a great impact on the field. Presently, most *in vivo* optical recordings are studies of neuronal or glial calcium dynamics. Neuronal calcium tracks action potentials as well as presynaptic and postsynaptic calcium signals at synapses, providing important information about both input and output signals. However, its ability to report subthreshold or inhibitory signals is variable, and while existing indicators have achieved single-spike sensitivity in low firing rate regimes, they cannot yet follow spikes in fast-spiking neurons.

The future of this field is not just improving calcium sensors, but generating a broad suite of optical sensors. Voltage indicators are ripe for development: by following voltage, one could in principle follow spikes and subthreshold signals, including inhibition. Several genetically encoded voltage indicators have appeared, but they do not yet have the desired combination of signal strength and speed, and could benefit greatly from disciplined, iterative improvements. Improved voltage indicators may well be genetically encoded, but other approaches from chemistry and nanotechnology should also be considered. The experience of optimizing the calcium indicators should be directly applicable to improving voltage indicators. Indicators with ultra-low background emissions hold particular importance for reliable event detection and timing estimation.

A major advance that could emerge from optical approaches is expanding the kinds of neuronal activity that are measured. For example, synaptic transmission is a rich area of research at the single neuron level, and could be accessible at the circuit level with better tools. Electrical synapses and their regulation are essentially invisible to most current recording methods. Direct measurements of released neurotransmitters at single-cell or single-synapse resolution are highly desirable: the probability of transmitter release at a synapse can vary 100-fold, and synapses also have properties such as depression and facilitation that shape signaling in real time. Existing methods for detecting transmitters such as voltammetry are useful but have limited spatial resolution. Direct measurement of released glutamate has recently been accomplished with a genetically-encoded sensor, offering potential improvement in both spatial and temporal resolution. Tools that allow direct measurements of other transmitters such as GABA, dopamine, serotonin, and neuropeptides would provide the needed view of synapses in action.

In the longer term, additional signaling properties may be monitored. For example, neuromodulatory states can dramatically change properties such as signaling dynamics, excitability, and plasticity. Measuring the biochemical readouts of neuromodulatory states (e.g. cAMP) may provide views of the slow processing that circuits perform in parallel to rapid computations. Glia are increasingly recognized as important players in neuronal signaling and pathology; monitoring glial activity and metabolic coupling may shed unexpected light into information processing in the brain. Monitoring synapses at a large scale could define the codes for transferring information across neuronal circuits and systems. A goal of particular interest is a way to find the synapses in a circuit that change as a result of experience and learning.

2c-iii. Integrated Optical Approaches: Neuroscience and Instrumentation

Optical methods capture the central vision of the BRAIN Initiative, that of integrating many approaches into a single experiment. Optical methods can be multiplexed to combine activity monitoring, manipulation, circuit reconstruction, and characterization of a single cell's morphology and molecular constituents (or at least a subset of the above) simultaneously. Similarly, combining electrode recording with optical methods provides added value. For example, including optical reporters like dyes or fluorescent proteins can help identify the recorded cell's identity and connectivity.

All of neuroscience will benefit from a streamlined integration of optical technologies for large-scale recording, optogenetic manipulation, and circuit reconstruction that allows multi-faceted studies of identified cells and circuits in individual brains. This will encompass unified development of compatible optical hardware, genetic or chemical activity reporters, and optogenetic tools. Technology for optical studies of brain dynamics and circuitry, cell types, and molecular content should be progressively developed over the long-term to attain sufficient throughput for sophisticated studies of the differences between individual subjects, in animals and in humans.

To reach their potential, optical methods should be viewed holistically. Wavelength ranges used for next-generation multi-color optical imaging and optogenetic control should ideally be tuned for mutual compatibility. Likewise, the capabilities and limitations of optical hardware should be taken into consideration when developing new sensor molecules, and *vice versa*, since the collective optical system is what ultimately should be optimized. For example, in the domain of optical sensors, much work is done at the surface of brain structures because imaging deep tissues remains a problem. Red or near infrared optical indicators would improve imaging depths in scattering tissues, but complementary strategies to solve this problem may be developed at the hardware-sensor interface, for example via nonlinear optical excitation using long wavelength illumination.

Optical engineering and photonics are rapidly progressing fields; ongoing advances in optical hardware and computational optics are likely to be highly pertinent to the BRAIN Initiative. Recent progress in miniaturized optics and CMOS image sensor chips for mobile phones has already yielded new capabilities for fluorescence imaging of neural activity in freely behaving animals. However, most emerging optical components will not have been tailored for neuroscience applications, systems engineering of new instrumentation using these components should pay careful heed to the unique needs of neuroscience experimentation.

Great benefit could come from short-term and sustained efforts to develop new instrumentation to improve the speed, tissue volume, tissue depth, and number of brain regions that can be monitored in live animals. These advances might come in many forms, such as: new hardware for high-speed imaging; parallelized detection systems; progress in miniature optics; novel light sources; microscopes with capabilities for large-scale recordings; wireless or automated imaging instrumentation; next-generation optical needles for imaging deep tissues; flexible optoelectronics; holographic or light-field techniques for precise optical interrogations in all three spatial dimensions; CMOS image sensors of larger size, finer pixels or built-in capabilities for image processing and automated detection of neural activity; or optical systems with scalable architectures and automated analytics for imaging in multiple animals or brain areas concurrently. Many of these instruments might exhibit both optical recording and manipulation capabilities.

Concurrent with the emergence of integrated optical approaches, it is essential to develop computational approaches for the analysis and management of the enormous data sets the optical techniques will yield (see also section 5). Calcium imaging studies in mice produce ~1 Gbits/sec of data; anatomical datasets will readily grow to the ~10 Petabyte scale and beyond. Sustained efforts will be necessary to develop sophisticated analytic tools for the analysis of these experiments. Policies and methods for data sharing will also need to be developed to fully exploit the value of these datasets (see section 8d).

At a deeper level, the concepts of optical imaging should be considered across other modalities such as magnetic fields or ultrasound. The value of existing technologies for human neuroscience, such as fMRI and MEG, is immense; developing higher-resolution methods for human use is an aspiration heard across the field. A non-invasive or minimally invasive imaging modality with cellular resolution that could interrogate large portions of the mammalian brain would represent a major advance for both animal and human studies. Any such technology that was safely applicable in humans would revolutionize our understanding of human brain function.

2c-iv. Nanotechnology and Unanticipated Innovations

As devices move from the micro- to nano-scale, properties emerge that may provide new opportunities to interrogate neurons. Silicon based nanodevices are one such example. Microwires, three microns in diameter, that project out from the surface of a conventional electrode can achieve intracellular access to cells plated over these wires; nanoposts, less than one micron in diameter, can create gigaohm seals with intracellular access for sustained periods of time. These devices have real promise; a high priority is to move their development from cell cultures to integrated neural systems, in slices or *in vivo*.

In the intermediate or long term, revolutionary new technologies may emerge, and the BRAIN Initiative should encourage their exploration and development. Nanodiamonds, for example, are particles whose photochemical properties—fluorescent light emission—might be tailored to exhibit sufficient sensitivity to applied electric fields to serve as optical reporters of electrical activity. With respect to neuroscience applications, however, essentially all aspects of this proposed technology are untested, from deployment of the particles within the neuronal plasma membrane to measurement of the light signals. DNA- and RNA-based technologies have been suggested as indirect reporters of neuronal activity that could be decoded by sequence.

These and other technologies should be encouraged and given "room to breathe" but held to a standard of progress: Emphasis should be placed on supporting methods and research teams that provide a logical, clear experimental pathway from an *in vitro* demonstration to a set of *in vivo* applications in increasingly complex neuronal systems. Nanoscale recording devices are potentially beneficial in terms of increased recording density, decreased tissue damage, and for long-term intracellular recording. However, they are all in very early stages, and moving this technology from *in vitro* to *in vivo* application is a significant challenge that will require sustained interaction between nanoscientists and neuroscientists.

High Priority Research Area for FY2014: Develop New Large-Scale Network Recording Capabilities. We should seize the challenge of recording dynamic neuronal activity from complete neural networks, over long periods, in all areas of the brain. There are promising opportunities both for improving existing technologies and for developing entirely new technologies for neuronal recording, including methods based on electrodes, optics, molecular genetics, and nanoscience, and encompassing different facets of brain activity, in animals and in some cases in humans.

3. Manipulating Circuit Activity

Observing natural patterns of neural activity generates hypotheses about their functional significance, but causal tests of such hypotheses require direct manipulation of the underlying neural activity patterns. In the 1950s, Penfield's electrical stimulation experiments suggested that a memory or thought could be elicited by activating neurons in the underlying network. In intervening years, electrical, chemical, and genetic methods for stimulating or inhibiting neurons have provided numerous insights. Currently, stimulating electrodes are being placed in human patients for spinal cord stimulation and deep brain stimulation, among other therapeutics. Despite these successes, current human stimulation methods lack precision and specificity, and could benefit from technological advances. In non-human neuroscience, a major recent advance in circuit manipulation has been the development of optogenetic tools based on light-activated channels and pumps. The combination of rapid activation, reliable effects, and genetic delivery of the optogenetic channels to specific cell types and brain regions has revolutionized modern neuroscience. Optogenetic tools for depolarizing and hyperpolarizing neurons have proved to be a general method for testing and generating hypotheses of brain function across systems, brain regions, and (non-human) species.

The existing optogenetic tools, generally based on Channelrhodopsin (depolarizing) and Halorhodopsin or Archaerhodopsin (hyperpolarizing), have been transformative but are not perfect for all uses. They have been subjected to multiple rounds of genetic engineering for optimization for different purposes, improvements demonstrating the importance of iteration in tool development. Nonetheless, new advances could make them still more useful. In the present versions, the light-induced currents are generally small, and the blue/ultraviolet light that most of these tools prefer is toxic to biological tissues, and does not penetrate deep into tissues. Most of the tools have overlapping absorption spectra with each other and with the genetically-encoded sensors of neural activity; they would be more useful if the spectra were separate. As mentioned earlier, improvements in optical physics may provide benefits for existing classes of tools, for example by allowing stimulation in complex, rapidly evolving patterns that imitate measured natural patterns of circuit activity. These kinds of improvements are incremental, but their cumulative impact could be substantial because these tools are so widely used.

There are broader possibilities for manipulating neuronal activity *in vivo*. Pharmacogenetic tools (such as RASSLs, DREADDs, and chemical-genetic switches for kinases and channels) are already a useful complement to optogenetics for long-term manipulation, and this is another area that will benefit from continued improvement. Entirely new tools could be developed based on magnetic stimulation, gases, infrared excitation, ultrasound, or organic or physical chemistry may allow access to neurons deep within the brain. Techniques of this sort could also allow independent access to multiple circuits, or independent tools for monitoring and manipulating neurons. Noninvasive and non-genetic approaches will be particularly important for human neuroscience.

There is also substantial room for growth in modulating more subtle aspects of neuronal function, not just depolarization and hyperpolarization. Among the tools that could have enormous impact are tools for silencing or activating particular synapses; affecting neuropeptide release independently of neurotransmitter release; or activating or inhibiting second-messenger cascades in real time, including those that mediate growth factor and neuromodulatory signals.

High Priority Research Area for FY2014: Develop A Suite of Tools for Circuit Manipulation. By directly activating and inhibiting populations of neurons, neuroscience is progressing from observation to causation, and much more is possible. To enable the immense potential of circuit manipulation, a new generation of tools for optogenetics, pharmacogenetics, and biochemical and electromagnetic modulation should be developed for use in animals and eventually in human patients. Emphasis should be placed on achieving modulation of circuits in patterns that mimic natural activity.

4. The Importance of Behavior

How can we discern the meaning of the complex, dynamic activity patterns in the brain? Neurologists often gain insight into a human brain disorder by observing a person's behavior, supplemented by person's verbal reports. For example, by measuring the behavior of the patient HM on objective tests, and by interacting with him on numerous occasions, the neuropsychologist Brenda Milner was able to demonstrate that his damaged hippocampus prevented him from forming explicit memories of events, but not implicit memories or habits. In non-human animals, however, we must rely exclusively on behavioral observation and measurements to gain insight into their cognitive processes. To understand an animal's perception, cognition, or emotion, we must start by observing its actions.

Applying next-generation recording and manipulation tools to evaluate behavior insightfully will become increasingly important to neuroscience. Behavioral metrics are most useful if they are objective and reliable, but they should also permit the study of rich behaviors appropriate to the species. Among existing behavioral methods, formal psychophysics has been powerful because of its use of detection tasks and choice (discrimination) tasks, which are standardized, quantifiable, and easily related to theoretical models such as signal detection theory. An alternative set of methods based on neuroethology examines freely-moving animals in naturalistic environments, where the behaviors for which the animal has evolved can be expressed. These experiments are typically combined with recordings from small implanted recording devices, with flexible tethers to allow free movement. A newer set of methods restrains the animal partially by holding its head fixed, but allows it to move on a tracking ball, giving it the perception of movement. These "fly on a ball" and "mouse on a ball" experiments can be combined with virtual reality environments in a closed-loop configuration, in which the animal's behavioral choices result in apparent changes in its environment; importantly, they permit the simultaneous use of complex optical or electrophysiological recording systems.

Advancing techniques for manipulating, tracking and analyzing animal behavior will be crucial components of recording and optogenetic experiments. Since no two animals have identical brains, matching neuronal and behavioral dynamics will be best achieved by conducting behavioral experiments over long durations in a single individual. Given the capabilities for tracking individual cells over many weeks in the living brain, designs of behavioral assays should fully exploit these long-term capabilities to reveal how the brain supports learning and memory, how the brain is altered in disease states, and how it responds to therapeutic manipulations. Building this behavioral capability should be a priority for the BRAIN Initiative. Ultimately, neuronal recordings and manipulations should intelligently scan as much as possible of the animal's behavior and cognitive repertoire.

One avenue for further growth is a more detailed understanding of behavioral dynamics. This, in turn, requires a framework for capturing animal behaviors in video or audio format, segmenting and classifying them, and determining their duration and the transitions between them. In an advance over methods requiring expert human observers, an emerging area combines neuroscience with machine vision and machine learning to automate analysis of high-dimensional behavioral data from video and auditory recordings. Subsequent behavioral segmentation and classification can be specified by the scientist (supervised learning) or detected entirely by a computer (unsupervised learning). This automated quantification of behavior has many appealing features that support its further development. It provides a high level of objectivity and consistency; it is labor-saving, enabling high-throughput and high-content analysis of behavior; and it has the potential to uncover new behavioral patterns that have been overlooked by human observers.

High Priority Research Area for FY2014: Link Neuronal Activity to Behavior. The clever use of virtual reality, machine learning, and miniaturized recording devices has the potential to dramatically increase our understanding of how neuronal activity underlies cognition and behavior. This path can be enabled by developing technologies to quantify and interpret animal behavior, at high temporal and spatial resolution, reliably, objectively, over long periods of time, under a broad set of conditions, and in combination with concurrent measurement and manipulation of neuronal activity.

5. Theory, Modeling and Statistics Will Be Essential to Understanding the Brain

Large brain data sets are accumulating at an unprecedented rate that will accelerate over the next decade as the BRAIN Initiative gathers momentum. The goal of brain theory is to turn this knowledge into understanding, but this is a formidable task. Brains—even small ones—are dauntingly complex: information flows in parallel through many different circuits at once; different components of a single functional circuit may be distributed across many brain structures and be spatially intermixed with the components of other circuits; feedback signals from higher levels constantly modulate the activity within any given circuit; and neuromodulatory chemicals can rapidly alter the effective wiring of any circuit. In complex systems of this nature, our intuitions about how the activity of individual components (e.g.

atoms, genes, or neurons) relate to the behavior of a larger assembly (e.g. macromolecules, cells or brains) often fail, sometimes miserably. Inevitably, we must turn to theory, simulation, and sophisticated quantitative analyses in our search to understand the underlying mechanisms that bridge spatial and temporal scales, linking components and their interactions to the dynamic behavior of the intact system.

Theory, modeling and statistics play at least four key roles in our effort to understand brain dynamics and function. First, for complex, frequently counterintuitive systems like the brain, mathematical modeling and simulations can organize known data, assist in developing hypotheses about underlying mechanisms, make predictions, and thus assist in designing novel experiments to test the hypotheses. Second, confirmatory statistical analysis allows us to move in the inverse direction after data collection, using formal inference approaches to support or disprove a stated theory or hypothesis. Third, exploratory data-mining techniques can be used to detect interesting regularities in complex data even when information cannot yet be summarized with the aid of forward modeling or when prior hypotheses do not yet exist. Data-mining techniques are powerful, but genuine understanding of underlying mechanisms will typically require subsequent confirmatory analyses of the usual hypothesis-testing variety. Finally, formal theory seeks to infer general principles of brain function that unify large bodies of experimental observations, models and simulation outcomes. The brain computes stably and reliably despite its construction from billions of elements that are both noisy, and constantly adapting and re-calibrating. Elucidation of the general principles underlying this remarkable ability will have a profound impact on neuroscience, as well as on engineering and computer science.

5a. Combining Theory, Modeling, Statistics, and Experiments

Theory and modeling have illuminated numerous areas of neuroscience in the past: the mechanisms of action potential generation (Hodgkin-Huxley), synaptic plasticity (Hebb), and visual motion computation (Hassenstein-Reichardt), the efficiency of sensory codes (Barlow), the role of inference and priors in perception (Helmholtz), the role of dopaminergic systems in computing prediction-errors for reinforcement learning (Schulz, Sutton, Dayan), and decision-making under uncertainty (Green and Swets, Luce). True partnerships between theorists and experimentalists will yield large dividends for almost every conceptual and experimental problem to be tackled under the BRAIN Initiative.

Modeling and theory developed through close collaborations of theorists and experimentalists are most likely to yield penetrating insight and drive creative experimental work. Data gathered by an experimentalist uninformed by theory, even excellent quality data, may not be the data that will generate the most definitive conclusions or greatest conceptual clarity. Similarly, theorists who participate actively in the acquisition of data are more likely to acquire a biological sense of the system, its reliability, and its limitations, increasing the likelihood that their theories reflect biological reality and make predictions that are feasible for experimental verification.

Ideally, theorists and statisticians should be involved in experimental design and data acquisition, not just recruited at the step of data interpretation. The close working relationships we envision could be supported by grant opportunities that require the participation of a statistician or theorist in collaboration with an experimentalist; a highly successful model for this has been the NIH-NSF CRCNS program.

We next highlight a few of the many areas that appear promising for the collaborative efforts of theorists and experimentalists under the goals of the BRAIN Initiative.

5a-i. New Statistical and Quantitative Approaches to New Kinds of Data

The next generation of neural activity recordings will be different from previous ones. All signs point to a major increase in the quantity of neuronal recordings, but the quality of neuronal recordings will also change during the BRAIN Initiative. Electrophysiological techniques are being complemented increasingly by optical recording methods, which will dictate different analytic approaches. The ability to identify specific cell types, and map this information onto activity maps, will provide an additional dimension to these rich data sets. The ability to record multiple forms of activity simultaneously – spiking, subthreshold activity, synchrony, neuromodulatory states – will increase as well. We will need new tools to analyze these complex datasets, as well as new tools and algorithms for data acquisition and interpretation: e.g., 1) different types of measurement will need to be fused to extract meaningful information, 2) the sheer amount of data will demand highly efficient algorithms, 3) in some cases, analyses will need to be done in real time, either because data volumes are too large for storage, or because the experiment is designed to adapt to the responses, or for applications such as neural prosthetics.

As new kinds of data become available through advances in molecular sensors and optical recording, equal effort must be expended to extract maximum insight from these novel data sets. Data analytic and theoretical problems are likely to emerge that we cannot anticipate at the present time. Resources should be available for experts from essential disciplines such as statistics, optimization, signal processing and machine learning to develop new approaches to identifying and analyzing the relevant signals.

5a-ii. Dimensionality and Dynamics in Large-Scale Population Recordings

In large-scale population recordings like those envisioned under the BRAIN Initiative, the issue of dimensionality is critical. Dimensionality reduction techniques detect correlated activity among subsets of the sampled neuronal population, identifying ensembles of neurons that might be functionally related to each other in interesting ways. In current recordings of tens-to-hundreds of neurons, the dimensionality of the data is typically much lower than the number of recorded neurons, encouraging the notion that different behavioral variables or neural circuitry constraints might be reflected in the activity of neurons in different ensembles. Low dimensionality is also interesting because it implies that one can capture the major sources of variation in a system by recording from a relatively small proportion of its neurons. Plainly, the issue of data dimensionality has substantial implications for what our experimental goals should be in large-scale population recordings. In a low-dimensional system, it would be far more important to sample neuronal activity strategically than to record from every neuron in the system.

There is reason for caution, however. The number of apparent dimensions in the data may be artificially low if the behavioral task is too simple or if neuronal activity is not measured for sufficiently long periods of time. Thus it is essential not only to record more neurons, but to also increase behavioral and stimulus complexity in order to obtain richer, higher dimensional data sets. A second problem is that many of these methods, such as ICA, PCA, support vector machines and graphical models, are designed for problems in which the structure in the data is static. More sophisticated statistical methods exist to analyze time series, and should be further developed to analyze the highly nonlinear dynamic structure of most neuroscience data.

The dynamics of neural activity in single cells, small circuits, and large populations are complex, and include nonstationary calibration, adaptation, and learning mechanisms that occur simultaneously and on different time scales. These are currently not well understood, and will require development of new theoretical and analysis tools based on control theory, information theory, and nonlinear dynamical systems.

Resolving the theoretical issues associated with dimensionality and dynamics, and developing new techniques and complex behavioral paradigms that can release the potential power of large neural data sets, are important contributions that theory and modeling can make to the BRAIN Initiative.

5a-iii. Linking Activity Across Spatial and Temporal Scales

One of the most remarkable properties of nervous systems is that the temporally-enduring behavior of an organism emerges from the collective action of molecules and cells operating on time scales many orders of magnitude shorter. At the cellular level, information is encoded in the patterns of action potentials generated by individual neurons, each enduring for roughly a millisecond. Yet a working memory may last tens of seconds, and a single purposeful behavior can extend for minutes or hours.

The integration of information across temporal scales is a problem that will involve biochemical signaling pathways that outlast an electrical input, circuit properties such as 'attractors', in which a large population of interacting nerve cells achieves an enduring activity state, and experience-dependent changes in the strength of synaptic connections between neurons. Meaningful models of brain activity will need to incorporate events at all of these temporal scales.

No less impressive is the extension of coordinated neural activity over large spatial scales within the brain. A purposeful behavior as simple as an eye movement can involve millions of neurons distributed across more than a dozen areas brain areas. How is widespread activity in the cortex, basal ganglia, thalamus, midbrain and brainstem orchestrated to achieve a single behavioral goal?

Theory, modeling and biophysically realistic simulations will play a critical role in deepening our understanding of these and many similar phenomena. Theory can help us develop hypotheses and design the right experiments to ask how purposeful systems-level behavior emerges at extended spatial and temporal scales.

5a-iv. Flexible Behavior and Decision-Making

Much of our knowledge of brain function comes from experimental measurements from one or a few neurons in a single brain area. As we attempt to understand the remarkably complex behavioral and cognitive abilities exhibited by humans, we will need to consider the interactions within and between larger neural systems and brain areas. Complex actions are driven by simultaneous inputs from multiple sensory modalities, as well as internal states and memories that represent goals, constraints, and preferences. These actions are readily adapted to different environments and contexts, and they can be learned and refined with experience. In addition to humans, many animals also demonstrate behaviors that can be flexibly reshaped or adapted according to context or task requirements. For example, neurons in the frontal lobes of non-human primates are centrally involved in decisions, and have been shown to respond to both task-relevant and task-irrelevant sensory stimuli, along with signals related to behavioral choices. The representation of these attributes within neuronal ensembles can change markedly during the execution of any given task. We do not yet have a systematic theory of how information is encoded in the chemical and electrical activity of neurons, how it is fused to determine behavior on short time scales, and how it is used to adapt, refine, and learn behaviors on longer time scales. Finally, humans and perhaps some animals have the capacity for symbolic computation using language and in other domains as well; a brain-based theory of these higher functions is notably lacking.

Coordinated work in theory, modeling and experiment will be required to understand the mechanisms of context-dependent information flow in the brain, which lies at the heart of flexible behaviors such as decision-making.

High Priority Research Area for FY2014: Integrate Theory, Modeling, Statistics, and Computation with Experimentation. Rigorous theory, modeling and statistics are advancing our understanding of complex, nonlinear brain functions where human intuition fails. New kinds of data are accruing at increasing rates, mandating new methods of data analysis and interpretation. To enable progress in theory and data analysis, we must foster collaborations between experimentalists and scientists from statistics, physics, mathematics, engineering and computer science.

6. Human Neuroscience and Neurotechnology

A primary goal of the BRAIN Initiative is to understand human brain function in a way that will translate new discoveries and technological advances into effective diagnosis, prevention, and treatment of human brain disorders. The study of human brain function faces major challenges because many experimental approaches applicable to laboratory animals cannot be deployed in humans. Nevertheless, direct study of the human brain is critical because of our unique cognitive abilities as well the profound personal and societal consequences of human brain disorders.

Improvements to existing technologies like magnetic resonance imaging (MRI) and positron emission tomography (PET) have revolutionized our ability to noninvasively study the structure, wiring, function and chemistry of the human brain. Other important opportunities are emerging from the increasing number of humans who are undergoing diagnostic brain monitoring with recording or stimulating electrodes, or are receiving neurotechnological devices for therapeutic applications or investigational studies (e.g. deep brain stimulation, or DBS). Some of the most promising new opportunities involve combining these and other techniques to cross barriers of spatial and temporal scale that have impeded progress in the past. For example, the inability to measure activity and chemistry at the cellular level with noninvasive tools creates significant uncertainty about the functional meaning of some of the recorded signals. We can potentially address this problem by combining noninvasive brain imaging with higher resolution data obtained from diagnostic monitoring of humans or from implanted devices in humans. Insight is also emerging by combined measurement of noninvasive and cellular-level signals in animal models. Breaking these barriers of scale (section 6a-ii) would yield substantial benefits for diagnosis and treatment of disease as well as for basic discovery about the human brain. These and other creative approaches for understanding human brain function should receive vigorous support under the BRAIN Initiative.

6a. Human Brain Imaging

The last twenty years have seen explosive growth in the development and use of noninvasive brain mapping methods, predominantly MRI, complemented by MEG and EEG, to investigate the human brain under normal and pathological conditions, and across the human lifespan. In the future, we anticipate significant progress in using these methods to measure the wiring diagram and functional activity of the human brain at multiple scales—neuronal ensembles, circuits and larger scale networks ('circuits of circuits'). In turn, these capabilities will allow us to visualize and understand circuit-level disruptions related to human brain diseases. Brain imaging techniques are also valuable for evaluating the effects of pharmacological treatments and non-invasive brain stimulation methods, or for validating other functional measurement methods like near-infrared spectroscopy.

6a-i. MRI Approaches

MRI techniques contribute extensively to human neuroscience in three broad ways: fMRI enables correlation of functional brain activity with cognition and behavior; diffusion-weighted

MRI (DW MRI) provides estimates of the trajectories of long-distance pathways in the white matter; resting-state fMRI (rfMRI) enables us to deduce 'functional connectivity' between dispersed brain regions. All of these investigations will be accelerated greatly if the spatial and temporal sampling limitations of the MR measurements can be reduced. The vast majority of current MR studies aim for whole-brain coverage and achieve spatial resolution of 2 mm (isotropic) or coarser. An 8 milliliter voxel from such an image contains a veritable world of smaller circuit components—more than 600,000 neurons and glial cells, many columnar ensembles, intermixed cortical layers, or (in the white matter) several fasciculated fiber bundles that may cross each other, fan out or turn within that volume.

The problem of relating MR signals to the underlying circuitry can be ameliorated somewhat by improved physical measurements. For example, in precisely targeted experiments using high field-strength magnets, hemodynamically based MR measurements have reached spatial scales below 1mm—to the level of cortical columns and individual cortical laminae. Even at this scale, however, MR measurements reflect a complex combination, at the vascular level, of electrochemical activity of many thousands of neurons and glia. Thus it is critically important to develop and experimentally validate theories of how MR signals are based in the underlying cellular-level activity, as we consider in the next section.

6a-ii. Bridging Spatial Scales

To link the integrative functional signals measured by MRI to cellular-resolution activity two separate issues warrant intensive investigation. First we must firmly establish how the electrical and chemical activity in different populations of excitatory and inhibitory neurons, glial cells, axons and presynaptic terminals contribute to the local vascular response—the classic neuro-glio-vascular coupling problem. This knowledge is immediately relevant not only for interpretation of fMRI signals, but also for investigation of neurological and psychiatric diseases in which a disruption in neuroglial communication and/or deterioration of neurovascular coupling contributes to motor and cognitive decline.

The second issue, which has received much less attention, concerns what information coded in the neural activity of populations of neurons within the imaged voxel is captured in the hemodynamic response. By analogy to bridging scales in physics, the basic intellectual question is how the electrical and chemical activity of neurons, glia and synapses in imaged voxels are integrated or averaged to generate a hemodynamic signal. These questions can be best answered by direct comparison of cellular-resolution population activity of neurons with hemodynamic response measurements, ideally in animals engaged in sophisticated behavior likely to invoke the full computational power of neural circuits. Additional approaches include the use of optogenetics, pharmacology, and mutant animals to perturb neuronal signals and directly observe the effects at the level of fMRI. These approaches offer the opportunity to bridge anatomical and physiological scales, going from cellular-resolution neuronal and glial activity to macroscopic circuits, networks and ultimately behavior.

DW MRI estimates the orientation of axonal fiber bundles, capitalizing on the fact that water diffuses most rapidly along the length of axons. By probing at many different orientations, DW

MRI can estimate not only the dominant fiber orientation in each voxel, but also the orientation of crossing fiber bundles, which are very common in the white matter thicket. Tractography algorithms combine information across a succession of voxels to estimate the overall trajectory of long distance pathways. Some of the problems faced by DW MRI in charting anatomical connectivity in the human brain, such as crossing fibers, fiber fanning, and so forth, may be alleviated by higher spatial resolution and more sensitive imaging techniques. The benefits of improved resolution have already been demonstrated by results from the Human Connectome Project (HCP), but these achievements still fall short of what is needed. Our ability to infer structural connectivity patterns using DW MRI will be further improved by using anatomically informed priors that are based on accurate statistical models of the distribution of trajectories taken by axons and fiber bundles in white matter. Such information can be obtained from cutting-edge microscopy methods (e.g., optical coherence tomography, polarized light imaging, CLARITY) and used to improve the fidelity of DW MRI-based tractography methods, thus bridging microscopic and macroscopic scales at the anatomical level. It is critical that at each stage of the process of improvement, the overall validity of DW MRI-based tractography is demonstrated by direct comparison with anatomical tract tracing in animal models, including non-human primates.

Thus, specific opportunities for MRI technology under the BRAIN Initiative include submillimeter spatial resolution descriptions of neuronal activity, functional and structural connectivity, and network analysis in the human brain through advances in instrumentation, data acquisition and analysis techniques, and theoretical modeling linking activity with behavior.

6a-iii. Resting State fMRI and Brain Network Dynamics

Resting state fMRI (rfMRI) enables inferences to be made about functionally connected networks that may be widely dispersed within the brain. It relies on the observation that functionally related areas that are co-activated during performance of a task also exhibit correlated spontaneous fluctuations when subjects are simply "resting" in the MR scanner. Many large-scale correlated temporal patterns, referred to as resting state networks (RSNs) have been identified in this manner. RSNs persist during sleep and under anesthesia, and are consistent across subjects and to some extent across species. Importantly, RSNs display some degree of correspondence with anatomical connections, but the correspondence is far from perfect. RSNs appear to reflect functional coordination operating across multiple synapses within a circuit, providing information about network topology that is difficult to infer from anatomical maps alone.

Recent developments in significantly accelerating whole brain functional images (e.g. 2 mm isotropic resolution, whole brain images acquired in less than a second) have enabled major improvements in spatial and temporal sampling of resting state fluctuations, leading to major gains in statistical power to detect RSNs, increases in the number and granularity of network nodes, and an enhanced ability to analyze neural dynamics (e.g. detection of regions dynamically participating in different networks). These recent advances emphasize the importance of gains enabled by improving spatial and temporal resolution of the MRI data, and the need to push the technology further.

Changes in RSNs have been implicated as possible biomarkers for functional classification in several cognitive disorders, and have enormous potential for further development in this area. Whether such rfMRI derived networks and/or network dynamics can inform us about individual differences in psychiatric disorders or guide individualized therapies is yet to be determined, but it is one of the significant potential payoffs of applying improved MRI methods and analysis techniques to be developed under the BRAIN Initiative.

6b. EEG and MEG

Electroencephalography (EEG) and magnetoencephelography (MEG) provide a unique capability for noninvasive analysis of human brain activity with high temporal resolution. Numerous studies have demonstrated the merits of EEG/MEG for detecting neural correlates of a broad range of human cognitive processes as well as brain disorders such as epilepsy. The simplicity and mobility of EEG monitoring systems has facilitated the study of human brain signals in naturalistic settings.

EEG/MEG is limited in its spatial resolution. Localizing EEG signals to specific brain structures (source imaging) has benefitted significantly from the *a priori* anatomic constraints measured with structural MRI. This synergistic interaction highlights the value of combining data across imaging measurement modalities. Recent advances in source imaging have significantly improved localization of event-related brain activity in healthy human subjects and of interictal spikes in epilepsy patients. A significant challenge for the future is to develop advanced source imaging techniques that can map spontaneous brain activity, including RSNs in healthy subjects, as well as abnormal network connectivity associated with neurological or psychiatric disorders.

An important opportunity lies in integrating high temporal resolution EEG (and MEG) source imaging with high spatial resolution fMRI. Significant progress has been made to leverage the complementary nature of EEG and fMRI, which can be performed simultaneously in an MRI scanner. Challenges exist to better understand the correlation between BOLD signals and electrophysiological events via neurovascular coupling, technologies for high fidelity recordings of brain activity using EEG-fMRI, and techniques to enhance performance of EEG source imaging from simultaneously acquired fMRI data. It is noteworthy that MEG/EEG methods fail to record spiking activity (the output signals of most neurons); this scale is missing in non-invasive recordings. New methods for recording neuron spiking externally would have a very large impact.

6c. PET and Neurochemistry

Assessment of dynamic neurochemical and other molecular events has been relatively neglected in recent years, in large measure because the techniques are difficult compared to fMRI. While MR spectroscopy offers a view of some important intrinsic molecules, it has not yet shown chemical specificity for neurotransmitter/receptor interactions. Optical and MR-based imaging methods offer significant potential for molecular imaging in animal models using exogenous probes, but the translation of these methodologies to humans is not straightforward. In the short- to mid-term, nuclear techniques, including single photon emission computed tomography (SPECT) and principally positron emission tomography (PET), provide the best means to translate studies of neurotransmitters, receptors, and neuromodulators to humans.

Two principal challenges limit the role of these techniques today. The first is to exploit the potential for better use of existing PET tracers that target dozens of important neurotransmitter systems and their receptor subtypes. Within the libraries of compounds tested for therapeutic potency by the pharmaceutical industry lie hundreds of compounds awaiting evaluation of their potential as *imaging* agents. Public/private partnerships under the BRAIN Initiative could unlock this potential treasure trove of compounds, not as therapeutic agents (for which they were originally evaluated), but as compounds for discovery of receptor function. Second, while the principle of using PET to evaluate changes in receptor occupancy secondary to pharmacological or cognitive stimulation has been demonstrated, the means for dynamic assessment of neurochemical-specific brain activation, analogous to fMRI localization of "activation", is not yet in hand.

Significant progress is possible, however, in both short and longer terms. True dynamic assessment of receptor occupancy and metabolism, at spatial resolution approaching today's fMRI studies and temporal resolution of minutes, is a feasible mid-term goal with key receptor subtypes of the dopaminergic, serotonergic and glutamatergic neurotransmitter systems. In the longer term, the range of molecular targets and receptor subtypes amenable for study should steadily grow, tapping into the breadth of neurochemical expertise available through partnerships with pharmaceutical companies.

High Priority Research Area for FY2014: Delineate Mechanisms Underlying Human Imaging Technologies. We must improve spatial resolution and/or temporal sampling of human brain imaging techniques, and develop a better understanding of cellular mechanisms underlying commonly measured human brain signals (fMRI, Diffusion Weighted MRI (DW MRI), EEG, MEG, PET)—for example, by linking fMRI signals to cellular-resolution population activity of neurons and glia contained within the imaged voxel, or by linking DW MRI connectivity information to axonal anatomy. Understanding these links will permit more effective use of clinical tools for manipulating circuit activity, such as deep brain stimulation and transcranial magnetic stimulation.

6d. Devices for Monitoring and Stimulating the Human Brain

A new generation of medical devices for interfacing with the living human brain has been fueled by the merger of engineering advances with neuroscience discovery. Devices, some already in hand, are being used to monitor brain function, to diagnose and treat mood and movement disorders, and to restore sensory and motor functions lost following injury or disease. Thousands of humans are receiving these neurotechnologies in clinically approved or investigational applications. With their informed consent, these individuals provide an extraordinary opportunity for rigorous research on normal brain function, as well as on the effects of brain injury or disease. When coupled with non-invasive imaging there is a real opportunity to bridge scales from limited cellular to whole brain functional imaging methods.

The population of humans receiving recording or stimulating devices is large and growing. Most notably, deep brain stimulation (DBS) electrodes implanted in a specific basal ganglia circuit have helped to relieve more than a hundred thousand people of the rigidity, tremor and slow movements of Parkinson's disease. DBS is also widely and successfully employed in motor disorders such as dystonia and tremor, and there is reason to think that its potential use is much broader: promising results have appeared for DBS use in intractable depression, and it is being explored as a treatment for obsessive-compulsive disorder and even memory decline, which could have major public health implications. Another frontier is 'closed-loop' implanted systems in which data analysis is performed in real-time by a computer and used to generate future patterns of brain stimulation. For example, a sensor might detect an epileptic seizure in the early stages of its development and reduce or block it by stimulating the brain into quiescence.

In the sensing domain, brain-computer interfaces (BCIs), still early stage investigational devices, can enable people with paralysis to use their own brain signals to control assistive prosthetics like computers or robotic arms well enough to perform some everyday activities of living. In these individuals, chronically implanted multielectrode sensors provide unprecedented high-resolution recording over years. Developments in sensing and stimulation technology promise a series of new devices that will increase the quality of life and independence of individuals limited by a wide range of brain injuries or disorders. Each person with a device, who is willing to participate, becomes a possible research participant with potential to yield valuable data about brain function,

The availability of this large cohort of people with implanted technology opens the possibility not only to advance clinical care, but to carry out detailed studies that were barely conceivable a decade ago. During intraoperative mapping in epilepsy, researchers found neurons in the medial temporal lobe of a human patient that responded to pictures of individual actors or politicians – and also to the spelled-out name of the same person. These neurons provide a fascinating example of the encoding of categories or abstract concepts in the brain. Direct brain recordings during anesthesia have revealed characteristic transformations in brain activity as consciousness is lost. Chronic multielectrode array recordings from brain-computer interfaces in people with longstanding paralysis have shown how the motor cortex retains representations of the arm even years after a stroke, raising new questions about plasticity in the human brain. As these applications continue to expand, there will be an unprecedented opportunity to study human circuits, both by recording their activity and modulating their activity.

6e. Teams for Basic and Clinical Human Research

Taking advantage of the scientific opportunities offered by neurotechnology developments and volunteer human patients is an exceptionally complex endeavor. Every opportunity should be maximized while maintaining the highest standards for research participant safety and protection. Meeting research goals and human research standards requires closely integrated research teams including clinicians, engineers, and scientists who work together to organize and carry out research of the highest integrity and rigor. Clinicians who use new neurotechnologies in human research should interact closely with the engineers, scientists and companies developing them, to ensure the rapid creation, validation, and dissemination of effective tools. The regulatory oversight of human research, as well as the close clinical relationship and potentially long-term commitment to participants, places a special time and financial burden on investigators. In addition, storing and processing data in compliance with federal privacy protection laws are challenging. Because related clinical research activities can occur across many universities and medical centers, mechanisms to standardize and share precious data from human subjects are essential. Further, because research is often aimed at creating new medical devices for treatment of human disease or injury, the objectives of the research team must often be aligned with regulatory paths and industry standards needed to translate early stage testing into a commercially viable technology. These varied demands must be met while also adhering to the highest standards of scientific quality and design of preclinical studies.

These standards can and are being met by dedicated teams of collaborative researchers, but consideration should be given to reducing unnecessary bureaucratic hurdles in the academic setting. This difficult but extremely valuable area of research should be made as efficient as possible. Initiating such steps could accelerate innovative research, ultimately driving down costs and providing better clinical devices and therapeutic outcomes to patients. Because this kind of research is so valuable yet so complex, we must develop novel incentives and straightforward mechanisms to translate basic science advances into human pilot testing, while maintaining the ethical standards and regulatory procedures in place for human research.

High Priority Research Area for FY2014: Create Mechanisms to Enable Collection of Human Data. Humans who are undergoing diagnostic brain monitoring or receiving neurotechnology for clinical applications provide an extraordinary opportunity for scientific research. This setting enables research on human brain function, the mechanisms of human brain disorders, the effect of therapy, and the value of diagnostics. Meeting this opportunity requires closely integrated research teams including clinicians, engineers, and scientists, all performing according to the highest ethical standards of clinical care and research. New mechanisms are needed to maximize the collection of this priceless information and ensure that it benefits people with brain disorders.

6f. Human Neural Technology Development

For human research, The BRAIN Initiative should ultimately support two broad types of technology development: (1) research tools that allow us to better investigate brain structure and function, and (2) clinical tools that enable us to better diagnose, prevent, treat, and cure brain diseases, including technologies that can restore lost functions. Devices for use in humans need substantial improvement over existing technology: they need to be more reliable, stable, and long lasting, which will require better materials, biocompatibility, and features optimized for human use. We need electrode arrays with higher spatial resolution for recording and stimulation both within and across brain areas. In the medium- to long-term, new monitoring capabilities (acoustic, optical, chemical, etc) should be incorporated into all implanted devices; when devices are implanted into human subjects, they should deliver the maximal scientific benefit consistent with health and safety of the subject. As detailed in section 2c-i, implantable devices must get smarter, smaller, and more energy efficient; they require wireless communication in compact packaging able to last for years in the body. Engineers and scientists must rise to the challenge of developing this next generation of neurotechnological devices.

Both penetrating and surface sensors need to be substantially improved and tested for their full capabilities. Precision in placing sensors within identified circuits will require MRI compatibility, which will also provide an opportunity to advance basic knowledge that links MRI identified circuits with clinical outcomes and cell and circuit scale function.

Potentially transformative technologies should also be entertained. For example, the use of optogenetics tools in humans is conceivable in the mid- to long-term. Initial safety studies of AAV (adenovirus-associated virus) vectors in human brains are encouraging, suggesting that viral delivery of therapeutic genes can be explored in the near future, with careful and comprehensive testing of viral delivery systems to evaluate long term safety and efficacy.

As previously mentioned, noninvasive tools for fine-resolution stimulation of the human brain would be transformative, potentially reducing or eliminating the need for invasive electrode implants. Present noninvasive stimulation techniques are being explored for therapeutic effects, including transcranial magnetic stimulation (TMS), and direct-current and slow alternating-current stimulation. These techniques are able to activate ~cm scale areas of brain for potential neurological and psychiatric applications. Frontal lobe TMS is already FDA approved to treat depression. However, the scale, duration, mechanism of action, and the potency of their effects need to be better elucidated. A long-term goal of the BRAIN Initiative should be to find ways to obtain high spatial and temporal resolution signal recording and stimulation from outside the head, perhaps through the use of minimally explored energy delivery techniques such as focused ultrasound or magnetic stimulation.

7. Education

New tools, whether they come in the form of equipment, molecular clones, or data analysis algorithms, should be disseminated to a wide scientific user base, along with the knowledge required to wield them.

7a. Education and Training in Emerging and Interdisciplinary Methods

New methods in molecular biology, optics, human brain imaging, and electrophysiology will require new training mechanisms. Funding training centers and personnel for teaching new techniques would require relatively modest funds and space, but would be a great benefit for the entire neuroscience community. Optogenetics has been successful in part because of organized mini-course training of faculty and students from around the world in the required surgeries and techniques, both in university settings and in course modules at Cold Spring Harbor and Woods Hole. Mini-courses in new technologies represent a way to bring an entire community of users up to a high level of understanding—and productivity—in a short period of time. They level the playing field between scientists at large institutions and those at smaller institutions who may not have the same resources. These teaching mechanisms have the added benefit of communicating experimental standards and pitfalls, which often trip up early users of new technologies, but currently suffer from a lack of standardized space and funding support in the traditional academic setting.

Training in quantitative neuroscience should be an area of special focus for the BRAIN Initiative. This includes teaching theory and statistics to biologists, and exposing physicists, engineers and statisticians to neuroscience. Mechanisms include fellowships as well as short courses and workshops in neuroinformatics, statistics, and computational neuroscience.

7b. Building Strength in Quantitative Neuroscience

Attracting new investigators to neuroscience from the quantitative disciplines (physics, statistics, computer sciences, mathematics, and engineering), and training graduate students and postdocs in quantitative neuroscience, should be high priority goals for the BRAIN Initiative. While forward-looking programs like the Sloan-Swartz Centers in Theoretical Neuroscience have attracted trainees from the quantitative disciplines and promoted their careers, this critical human asset remains too small and too tenuously established within neuroscience. Too many neuroscience departments remain skeptical of hiring faculty whose research does not focus primarily on experimental lab work, and too many statistics, physics, mathematics and engineering departments are hesitant to hire faculty who focus intensively on the nervous system.

The field would benefit greatly from incentives for faculty recruitment at this critical interface. A major benefit of attracting new faculty from the quantitative disciplines to neuroscience would be their teaching of quantitative concepts and skills—theory, modeling, statistics, signal processing, and engineering in its many forms—with emphasis on real-world applications to neuroscience data sets, including those introduced to the public domain under BRAIN-sponsored research projects. Most students currently enter neuroscience graduate programs with little-to-no training in statistics, computer science, or mathematical modeling, and many receive little formal quantitative training during their graduate education. This must change. All areas of neuroscience, not just those of particular emphasis under the BRAIN Initiative, will become increasingly dependent on quantitative perspectives and analyses in the future;

training of students in quantitative reasoning, principles, and techniques must increase accordingly.

High Priority Research Area for FY2014: Disseminate Knowledge and Training. Progress would be dramatically accelerated by the rapid dissemination of skills across the community. To enable the broadest possible impact of newly developed methods, and their rigorous application, support should be provided for training—for example, summer courses and course modules in computational neuroscience, statistics, imaging, electrophysiology, and optogenetics—and for educating non-neuroscientists in neuroscience.

8. Maximizing the Value of the BRAIN Initiative: Core Principles

The emphasis in this interim report is on posing questions, not dictating solutions. However, certain principles and approaches can maximize the intellectual value and long-term impact of all aspects of the BRAIN Initiative, as enumerated below.

8a. Use Appropriate Experimental Systems and Models

Our ultimate goal is to understand the human brain, and as stated above, human neuroscience should be a key element of the BRAIN Initiative. However, both ethical principles and scientific feasibility will require many methods and ideas to be developed in non-human animal models, and only later applied to humans. With a few exceptions, we do not emphasize particular animal models but instead wish to encourage a diversity of approaches. The history of neuroscience teaches us that many different animal models should be enlisted for the unique advantages that they provide, and also that comparative approaches are very powerful in discovering biological principles. We expect the BRAIN Initiative to include nonhuman primates such as rhesus macaques, because they are evolutionarily the closest animal model for humans, and this will be reflected in their behavioral and cognitive abilities, genetics, anatomy, and physiology. We expect the mouse to be the initial mammalian model for the use of genetic tools, supplemented by the rat, long appreciated for its behavior and neurophysiology, where genetic tools are also emerging. The transparent zebrafish larva should facilitate optical recording methods in the context of a simplified vertebrate neuroanatomy. Invertebrate animals with smaller nervous systems offer rapid experimental turnaround, rapid testing and validation of new tools, and the ease of genetics (for worms and flies) or of electrophysiology (for molluscs, crabs, and leeches) targeted to defined neurons; most neuroscientists have been surprised to see how many features of the brain and behavior are shared by vertebrates and invertebrates.

Finally, the list of species above is not complete. It is important to realize how much has been gained from studying a wider variety of animal species, recognizing their special abilities and the perspective they provide on the brain. For example, the only animals for which a teacher instructs vocal learning, other than humans, are songbirds. The richness of the behavioral repertoire in songbirds has led to remarkable insights into learning, motor control, and the importance of social context in behavior. Important insights into the brain have come from studies of many other creatures (barn owls, electric fish, chickens, bats, and more). The most significant technologies developed by the BRAIN Initiative should facilitate experiments in these and other specialized animals, broadening the reach and scope of questions that can be asked about the brain.

The fundamental principle is that experimental systems should be chosen based on their power to address the specific question at hand. Although the emphasis of the BRAIN Initiative is on the whole brain, some technologies will require careful analysis in culture systems or slices before they can be used in intact animals or humans. The BRAIN Initiative should not be dogmatic when faced with a compelling scientific argument for a different approach.

8b. Cross Boundaries in Interdisciplinary Collaborations

Surveying the landscape of neurotechnologies reveals some that are mature, some that are emerging and in need of iterated, disciplined improvement, and some that require re-imagination. It is critical that the BRAIN Initiative boldly supports the very best ideas addressing each need. This report describes the current state of the field, but transformative new ideas will emerge in the future that are not on today's horizon. The BRAIN Initiative must find a way to recognize such new ideas and let them flourish. Some innovative ideas will certainly fail, but this is not the time to play it safe. If the majority of proposals succeed in a predictable manner, we are not being adventurous enough.

At this stage, it is senseless to choose a single funding mechanism or set of investigators for the BRAIN Initiative; there must be exploration. Applications should be solicited widely, with open competition for resources. Some ideas will be initiated by individual investigators who see a new way forward. Other ideas will require larger teams of scientists, particularly in human neuroscience with its unique ethical and scientific challenges.

A theme that emerged clearly from the working group's workshops and discussions is the benefit to be gained by new scientific partnerships that cross traditional areas of expertise. This point was made in many specific contexts:

> The physicists and engineers who develop optical hardware should partner with the biologists and chemists who develop new molecular sensors.

> The tool-builders who design new molecules for sensing or regulating neurons should partner with neuroscientists who will rigorously examine their validity in neurons and brains.

> The theorists who develop models for understanding neuronal dynamics should partner with experimentalists, from initial experimental design to execution to interpretation.

The clinicians and neuroscientists who develop sophisticated imaging methods in humans should partner with scientists working in animal models who can relate imaging signals to the underlying cellular mechanisms with great precision.

Supporting collaborations across disciplines, with outstanding scientists who are intellectual equals, could light new fires in technology development. Such groups need not be at one institution to be effective – the quantitative and physical scientists might be at engineering schools, their neuroscientist partners might be at medical schools. Small collaborating groups of two or three investigators could open new doors in ways that no single investigator or conventional department would imagine; the BRAIN Initiative should particularly stimulate this kind of partnership.

8c. Integrate Spatial and Temporal Scales, and Accelerate All of Neuroscience

As mandated by the charge to our working group, this report focuses on new research opportunities at a critical level of neuroscience investigation—that of circuits and systems. As described in detail in this report, however, circuits and systems cannot be understood incisively without reference to their underlying components—molecules, cells and synapses. Neither can circuits and systems be understood without reference to the whole brain, the behavior of the organism, and how brain circuits are shaped by the unique experiences of the individual. The brain must be understood as a mosaic unity encompassing all of these levels.

The particular focus of the BRAIN Initiative represents only one important aspect of neuroscience, but one that can benefit many other areas. The funds devoted to the BRAIN Initiative are a very small fraction of the NIH's total investment in neuroscience and neurological disorders. To have maximal impact, the new knowledge and technology created under BRAIN must focus, but its products must accelerate all other subdisciplines of neuroscience so that they also advance and flourish.

Appropriately, a substantial fraction of the NIH's investment in neuroscience is allocated to specific human brain disorders. The recommendations in this report have been composed with a specific eye toward their eventual impact for humans—in translational neuroscience research, in medical practice to alleviate suffering (some technologies, such as brain imaging and stimulation, are already in widespread use in medicine), and in other areas such as education. In the nearer term, methods from the BRAIN Initiative can be applied in animal models of human brain disorders, seeking insight about fundamental disease mechanisms and asking about the sources of variation in neurological function across individuals.

Circuit-level studies must be synergistic with disease-related studies at other levels. In the field of human genetics, for example, progress is being made in understanding genetic risk factors that contribute to psychiatric and neurological diseases. Tools from the BRAIN Initiative, applied to human subjects and animal models, will allow researchers to ask how these genetic risk factors cascade upward through the properties of defined populations of neurons, circuits, brain regions, and cognitive processes. Another area in which great advances are occurring is stem cell biology, notably the development of induced pluripotential stem cells (iPS cells) from healthy individuals and patients that can be differentiated into mature neurons. By providing insight into the essential features of the brain, such as its constituent neurons, their connections, and their patterns of activity in health and disease, the BRAIN Initiative will provide a standard for evaluating the competence of iPS-derived neurons to form functional circuits, and a stimulus for improving them to the point that they can be considered as therapeutic tools.

There are many points of intersection between other areas of basic neuroscience and the BRAIN Initiative as well. A delineation of neuronal cell types and their patterns of gene expression should be a resource for cellular and molecular neuroscience. The census of cell types, and studies of their connectivity, should provide new tools to study central questions in developmental neuroscience. Technologies for real-time measurements of neuronal activity, neuromodulators, and synaptic connections are ideally suited for use in cellular and slice neurophysiology; indeed, many will be used there before they can be applied in whole animals.

8d. Establish Platforms for Sharing Data

The traditional way to exchange scientific information is through publications and books, but we have entered a new age of information that is not limited by the narrow bandwidth of journals. High-speed computing and massive storage capabilities have enabled collection of much larger datasets than was previously possible, and the Internet has enabled data sharing on a far wider scale. However, many datasets that are currently available are diverse, fragmented, and highly dynamic, which is to say unstable. Currently most of the raw data that go into published papers are not available outside the laboratory where they were collected. Inevitably, this leads to duplication of effort, inefficiency, and lost knowledge.

Well-curated, public data platforms with common data standards, seamless user accessibility, and central maintenance would make it possible to preserve, compare, and reanalyze valuable data sets that have been collected at great expense. This would be of great benefit to neuroscience, just as the availability of public genomic and protein structure databases have transformed genetics and biochemistry. Analysis tools and user interfaces should be developed that can be run remotely, such as the BLAST program for sequence alignment in genomic databases. Creating and maintaining such data platforms would entail a major effort of the community to decide what data and metadata to include, controls on the use of data, and support for users. Valuable lessons and best-practices can be learned from existing public datasets, which include the Allen Brain Atlas, the Mouse Connectome Project, the Open Connectome Project, the CRCNS data sharing project, ModelDB and the Human Connectome Project, as well as datasets generated by the physics, astronomy, climate science, and technology communities. A first unifying attempt, the Neuroscience Information Framework (NIF) sponsored by NIH, provides a portal to track and coordinate multiple sites, but the myriad genetic, anatomical, physiological, behavioral and computational datasets are difficult to manage because of their heterogeneous nature. The NIH Big Data to Knowledge Initiative offers opportunities to neuroscientists to develop new standards and approaches.

Methods and software as well as data should be shared. Some neural simulators such as Genesis, NEURON and MCell are well-established, open source and well documented, but the software for many models and simulations in published papers are undocumented or unavailable. The description of a model in a published paper is often insufficient to reproduce the simulations; it is essential that software be made available so that all models in published papers are reproducible. As data sets become larger and as new types of data become available, there is increasing need for public, validated methods for analyzing and presenting these data. As an example, microelectrode recordings often pick up spikes from several neurons that need to be separated into single units—a procedure known as "spike sorting". A plethora of custom spike-sorting programs have been created by many individual laboratories. But there are no widely accepted standards for rigorous spike sorting, and it can therefore be difficult to compare data precisely across laboratories. The community would benefit from common standards for spike-sorting and for other common data analysis procedures.

New data platforms would also encourage changes in the culture of neuroscience to promote increasing sharing of primary data and tools. We heard from many researchers about the value of sharing data, and their desire for stable, easily interconvertible data formats that could accelerate the field. Data and data analysis tools that emerge in the BRAIN Initiative should be freely shared to the extent possible, no later than the date of their first publication and in some cases prior to that date. Some areas of neuroscience, such as human brain imaging (the Human Connectome project; the International Neuroimaging Data-Sharing Initiative), are already sharing data on a large scale despite the enormous datasets involved. Having said that, extending this model to all fields is a difficult problem, and cannot be solved at one step. Based on the history of data sharing in many fields of biology, the solution will come from the engagement of sophisticated, motivated members of the scientific community from the bottom up, not from a directive from above.

To meet these goals, BRAIN Initiative will require infrastructure for integrating and sharing relevant datasets and data analysis methods. There is much that could be done in partnership with computer scientists and database experts to set standards for data formats and best practices for maintaining and disseminating heterogeneous data. The infrastructure for maintaining common databases will require dedicated resources, which may be provided by the NIH Big Data to Knowledge Initiative or the NIH Neuroscience Blueprint to support the BRAIN Initiative.

8e. Validate and Disseminate New Technology

A primary goal of the BRAIN Initiative will be to identify and support new technologies with potential to substantially accelerate high quality brain research. Technology development begins with innovation, but it is a continuing process. The first genetically-encoded fluorescent calcium indicator, cameleon, was published in 1997; the current versions, such as GCaMP6, were published in 2012 after a focused and sustained effort over ~5 years at a total cost of ~$10 million. The basis of the method has not changed, but its utility and applications have increased immensely.

This example and others show that technologies become valuable after they have gone through the processes of validation in biological systems (with comparison to the current best practices), iteration (serial improvements in properties), application (to a variety of test systems), and dissemination (including education and training). The entire neuroscience community would benefit from support for accelerated technology development between the initial proof-of-principle and the mature system; incisive new research could be accelerated by decades.

Following technology development, the BRAIN Initiative should build an infrastructure for sharing relevant tools of all kinds, whether biological, chemical, or physical. Molecular biology tools and viruses are easily disseminated, although there is some associated cost that must be supported. Other tools in discussion from chemistry, nanoscience, and physics are not so easily sent in the mail. If the BRAIN Initiative develops next-generation electrodes, nanotechnologies, or chemical probes, it should ensure that they can be synthesized, fabricated, or readily purchased by researchers, as appropriate. If the BRAIN Initiative funds development of a next-generation microscope, it should ensure that private or publicly supported mechanisms make it available to a variety of users, not just the inventors. A pathway to dissemination should be expected for BRAIN-derived tools, whether that is commercialization, core facilities, or something else. Computational and statistical tools developed under the BRAIN Initiative should also be supported and broadly available. Extremely complex technologies like next-generation high-field MRI instruments might need to be centralized, like the centralized X-ray beam lines used by structural biologists. Core facilities could be established with state-of-the-art technology, perhaps at a few universities or research institutes, but allowing use by researchers from across the country. Such core facilities could be attractive enough to the host institutions that they would co-invest in equipment and support personnel.

In summary, a core principle of the BRAIN Initiative is that new technologies and reagents should be made available across the community at the earliest possible time. This will require a thoughtful development of dissemination policies by the scientific community, as well as specialized support mechanisms, private-public partnerships, and training programs.

8f. Consider Ethical Implications of the BRAIN Initiative

The working group is extremely pleased that the President has charged his Bioethics Commission with exploring the ethical issues associated with the conduct of neuroscience research, and also the ethical issues surrounding the application of neuroscience research findings in medicine and other settings. Many ethical and policy issues raised by the BRAIN Initiative are not unique to neuroscience research, and thus we can learn from ongoing experiences in other fields. For example, the NIH BRAIN Working Group has endorsed data sharing, since advances in science are often catalyzed by collaborations and open access to data. However, our experiences with genetic data have shown that privacy concerns must be managed carefully to protect human research participants. Other issues such as defining what constitutes "enhancement" and how we obtain consent from potentially vulnerable populations have been debated widely across biomedical research, and will continue under the BRAIN Initiative.

Although brain research entails ethical issues that are common to other areas of biomedical science, it entails special ethical considerations as well. Because the brain gives rise to consciousness, our innermost thoughts and our most basic human needs, mechanistic studies of the brain have already resulted in new social and ethical questions. Can research on brain development be used to enhance cognitive development in our schools? Under what circumstances should mechanistic understanding of addiction and other neuropsychiatric disorders be used to judge accountability in our legal system? Can civil litigation involving damages for pain and suffering be informed by objective measurements of central pain states in the brain? Can studies of decision-making be legitimately used to tailor advertising campaigns and determine which products are more attractive to specific consumer bases? Brain research must proceed with sensitivity and wisdom. The working group looks forward to the deliberations of the Bioethics Commission, and to interacting with the group to establish a scientifically rigorous plan for the BRAIN Initiative that is grounded in sound ethical policies. As is clear from the scientific issues reviewed in this report, developing a deep understanding of the brain is only possible through research on animals and informed, volunteer human subjects. Without question, research under the BRAIN Initiative should adhere to the highest ethical standards for research with human subjects and with non-human animals, within the regulatory framework of the United States and host research institutions.

CONCLUDING REMARKS

In summary, as we present this interim report, we see immense potential in BRAIN Initiative technology applied to compelling neuroscience questions. We see enormous opportunities in providing new tools to basic neuroscientists, translational researchers, neurologists, psychiatrists, radiologists, and neurosurgeons. At the same time, we recognize that these initial goals will be refined, that new goals may emerge, and that objectives and priorities are still crystallizing. As we continue this planning process, we hope to represent the best collective scientific wisdom of the field. We look forward to the advice of our colleagues in neuroscience, medicine, psychology, biology, chemistry, and the quantitative sciences, and to the advice of patient advocates and the public, as we shape a mature vision and sharper goals for our final report to the ACD in June 2014. The BRAIN Initiative is a challenge and an opportunity to solve a central mystery—how organized circuits of cells interact dynamically to produce behavior and cognition, the essence of our mental lives.

The resources listed below introduce some of the neurotechnologies and BRAIN Initiative-related resources that are described in the report. Most papers are reviews, although a few recent methods papers are included. This is not a comprehensive citation list.

THE BRAIN INITIATIVE

Obama, BH www.whitehouse.gov/the-press-office/2013/04/02/remarks-president-brain-initiative-and-americaninnovation.

NIH BRAIN Initiative, http://www.nih.gov/science/brain/index.htm

Insel TR, Landis SC, Collins FS (2013) Research priorities. The NIH BRAIN initiative. Science 340(6133):687-688.

MAPPING THE STRUCTURE AND COMPONENTS OF CIRCUITS

Cell Type

Bernard A, Sorensen SA, Lein ES (2009) Shifting the paradigm: new approaches for characterizing and classifying neurons. Curr Opin Neurobiol 19(5):530-536.

Lein ES et al (2007) Genome-wide atlas of gene expression in the adult mouse brain. Nat 445(7124):168–176.

Experimental Access to Cell Types

Gaj T, Gersbach CA, Barbas CF 3rd. (2013) ZFN, TALEN, and CRISPR/Cas-based methods for genome engineering. Trends Biotechnol. 2013 31(7):397-405.

Jenett A et al (2012) A GAL4-driver line resource for Drosophila neurobiology. Cell Rep 2:991-1001.

Huang ZJ, Zeng H (2013). Genetic approaches to neural circuits in the mouse. Ann Rev Neurosci 36:183-215.

Structural Maps

Denk W, Briggman KL, Helmstaedter M (2012) Structural neurobiology: missing link to a mechanistic understanding of neural computation. Nat Rev Neurosci 13(5):351-8.

Ginger M, Haberl M, Conzelmann, KK, Schwarz, MK, Frick A (2013) Revealing the secrets of neuronal circuits with recombinant rabies virus technology. Frontiers in Neural Circuits 7(2):1-15.

Kleinfeld D et al (2011) Large-scale automated histology in the pursuit of connectomes. J Neurosci 31(45):16125-16138.

Osten P, Margrie TW (2013) Mapping brain circuitry with a light microscope. Nat Methods 10:515-523.

Human Connectome Project, http://www.humanconnectomeproject.org/

Mouse Connectome Project, http://www.mouseconnectome.org/

Clarity Resources, http://clarityresourcecenter.org

NEURONAL DYNAMICS: RECORDING NEURONAL ACTIVITY ACROSS TIME ANDS SPACE

Recording from Complete Circuits
Ahrens MB, Orger MB, Robson DN, Li JM, Keller PJ (2013) Whole-brain functional imaging at cellular resolution using light-sheet microscopy. Nat Methods10(5):413-420.

Alivisatos AP, Chun M, Church GM, Greenspan RJ, Roukes ML, Yuste R (2012) The brain activity map project and the challenge of functional connectomics. Neuron 74:970-974.

Advancing Recording Technology (Electrophysiology)
Buzáki G (2004) Large-scale recording of neuronal ensembles. Nat Neurosci 7(5):446-451.

Szuts TA et al (2011) A wireless multi-channel neural amplifier for freely moving animals. Nat Neurosci 14(2):263-270.

Advancing Recording Technology (Optical sensors)
Looger LL, Griesbeck O (2012) Genetically encoded neural activity indicators. Curr Opin Neurobiol 22(1):18-23.

Peterka DS, Takahashi H, Yuste R (2011) Imaging voltage in neurons. Neuron 69:9-21.

Integrated Optical Approaches: Neuroscience and Instrumentation
Wilt BA, Burns LD, Wei Ho ET, Ghosh KK, Mukamel EA, Schnitzer MJ (2009) Advances in light microscopy for neuroscience. Annu Rev Neurosci. 32:435-506.

Nanotechnology and Unanticipated Innovations
Alivisatos AP et al (2013) Nanotools for neuroscience and brain activity mapping. ACS Nano 7(3):1850-1866.

Spira ME, Hai A (2013) Multi-array technologies for neuroscience and cardiology. Nat Nanotechnol. 2013 8(2):83-94

MANIPULATING CIRCUIT ACTIVITY
Fenno L, Yizhar O, Deisseroth K (2011) The development and application of optogenetics. Annu Rev Neurosci 34:389-412.

Farrell MS, Roth BL (2013) Pharmacosynthetics: Reimagining the pharmacogenetic approach. Brain Res 1511:6-20.

Packer AM, Roska B, Hausser M (2013) Targeting neurons and photons for optogenetics. Nat Neurosci 16(7):805-815.

THE IMPORTANCE OF BEHAVIOR
Dombeck DA, Reiser MB (2012) Real neuroscience in virtual worlds. Curr Opin Neurobiol 22(1):3-10.

Kabra M, Robie AA, Rivera-Alba M, Branson S, Branson K (2013) JAABA: interactive machine learning for automatic annotation of animal behavior. Nat Methods 10(1):64–67.

THEORY, MODELING AND STATISTICS

Ganguli S, Sompolinsky H (2012) Compressed sensing, sparsity, and dimensionality in neuronal information processing and data analysis. Annu Rev Neurosci 35:485-508.

Marder E, Taylor AL (2011) Multiple models to capture the variability in biological neurons and networks. Nat Neurosci 14:133–138.

Shenoy KV, Sahani M, Churchland MM (2013) Cortical control of arm movements: a dynamical systems perspective. Annu Rev Neurosci 36:337-359.

Wang XJ (2013) The prefrontal cortex as a quintessential "cognitive-type" neural circuit: working memory and decision making. In: Principles of frontal lobe function, Second edition (Stuss DT, Knight RT, eds), pp. 226-248. New York: Oxford UP.

Kass RE, Ventura V, Brown EN (2009) Statistical issues in the analysis of neuronal data. *J Neurophys* 94(1):8-25.

HUMAN NEUROSCIENCE AND NEUROTECHNOLOGY

Donoghue JP (2008) Bridging the brain to the world: a perspective on neural interface systems. Neuron 60(3):511-521.

Fox MD, Halko MA, Eldaief MC, Pascual-Leone A (2012) Measuring and manipulating brain connectivity with resting state functional connectivity magnetic resonance imaging (fcMRI) and transcranial magnetic stimulating (TMS). Neuroimage 62:2232-2243.

Holtzheimer PE, Mayberg HS (2011). Deep brain stimulation for psychiatric disorders. Annu Rev Neurosci 34:289-307.

Lozano AM, Lipsman N (2013) Probing and regulating dysfunctional circuits using deep brain stimulation. Neuron 77(3):406-424.

McNab JA et al (2013) The Human Connectome Project and beyond: Initial applications of 300 mT/m gradients. Neuroimage 80:234-245.

Smith SM et al (2013) Resting-state fMRI in the Human Connectome Project. Neuroimage 80:144-168.

Sotiropoulos SN et al (2013) Advances in diffusion MRI acquisition and processing in the Human Connectome Project. Neuroimage 80:125-143.

PRINCIPLES

Data Platforms, Data Sharing, and Big Data

Akil H, Martone ME, Van Essen DC (2011) Challenges and opportunities in mining neuroscience data. Science 331(6018):708-712.

Berman F, Cerf V (2013) Science priorities. Who will pay for public access to research data? Science 341(6146):616-617.

Ng L et al (2009) An anatomic gene expression atlas of the adult mouse brain. Nat Neurosci 12(3):356–362.

ETHICAL CONSIDERATIONS

Presidential Commission for the Study of Bioethical Issues – BRAIN Initiative, http://bioethics.gov/node/2629

MAY 29, 2013 – MOLECULAR APPROACHES

Edward Callaway, PhD, Audrey Geisel Chair and Professor, Salk Institute for Biological Studies

Nathaniel Heintz, PhD, James and Marilyn Simons Professor, Rockefeller University

Ehud Isacoff, PhD, Professor and Head of Neurobiology, University of California, Berkeley

Loren Looger, PhD, Group Leader, Janelia Farm Research Campus

Liqun Luo, PhD, Professor of Biology, Stanford University

Clay Reid, PhD, Senior Investigator, Allen Institute for Brain Science

Gerald Rubin, PhD, Vice President and Executive Director, Janelia Farm Research Campus

Michael Stryker, PhD, WF Ganong Professor of Physiology, University of California, San Francisco

Alice Ting, PhD, Ellen Swallow Richards Associate Professor of Chemistry, Massachusetts Institute of Technology

Hongkui Zeng, PhD, Senior Director, Allen Institute for Brain Science

Feng Zhang, PhD, Core Member, Broad Institute; Investigator, McGovern Institute for Brain Research; and Assistant Professor of Neuroscience, Massachusetts Institute of Technology

JUNE 26, 2013 – LARGE-SCALE RECORDING TECHNOLOGIES AND STRUCTURAL NEUROBIOLOGY

Edward Boyden, PhD, Benesse Chair, New York Stem Cell Foundation-Robertson Investigator, and Paul Allen Distinguished Investigator, Massachusetts Institute of Technology; Associate Professor, MIT Media Lab and McGovern Institute

György Buzsáki, MD, PhD, FAAAS, Biggs Professor of Neural Sciences, New York University

Winfried Denk, PhD, Director, Max Planck Institute for Medical Research

Florian Engert, PhD, Professor of Molecular and Cellular Biology, Harvard University

Michale Fee, PhD, Professor, McGovern Institute, Massachusetts Institute of Technology

Jeff Lichtman, MD, PhD, Professor of Molecular and Cellular Biology, Harvard University

Markus Meister, PhD, Professor of Biology, California Institute of Technology

Pavel Osten, MD, PhD, Associate Professor, Cold Spring Harbor Laboratory

Hongkun Park, PhD, Professor of Chemistry and Physics, Harvard University

Kristin Scott, PhD, Associate Professor, University of California, Berkeley

Karel Svoboda, PhD, Group Leader, Janelia Farm Research Campus

Rafael Yuste, MD, PhD, Professor, Columbia University

JULY 29, 2013 –COMPUTATION, THEORY, AND BIG DATA

Kwabena Boahen, PhD, Professor of Bioengineering, Stanford University

Kristin Branson, PhD, Lab Head, Janelia Farm Research Campus

Jennifer Chayes, PhD, Distinguished Scientist and Managing Director, Microsoft Research

Todd Coleman, PhD, Associate Professor of Bioengineering, University of California, San Diego

Uri Eden, PhD, Associate Professor of Statistics, Boston University

Jack Gallant, PhD, Professor of Psychology, University of California, Berkeley

Surya Ganguli, PhD, Assistant Professor of Applied Physics, Stanford University

Stephanie Jones, PhD, Assistant Professor of Neuroscience, Brown University

Nancy Kopell, PhD, William Fairfield Warren Distinguished Professor and Co-Director of the Center for Computational Neuroscience and Neural Technology, Boston University

Maryann Martone, PhD, Professor-in-Residence, University of California, San Diego

Bruno Olshausen, PhD, Director of the Redwood Center for Theoretical Neuroscience and Professor, University of California, Berkeley

Patrick Purdon, PhD, Instructor of Anaesthesia, Harvard University, and Assistant in Bioengineering, Massachusetts General Hospital

Sebastian Seung, PhD, Professor of Computational Neuroscience, Massachusetts Institute of Technology

AUGUST 29, 2013 – HUMAN NEUROSCIENCE

Krystof Bankiewicz, MD, PhD, Professor of Neurosurgery and Neurology and Kinetics Foundation Chair in Translation Research, University of California, San Francisco

Sydney Cash, MD, PhD, Associate Professor of Neurology, Harvard University

Timothy Denison, PhD, Director of Core Technology and Technical Fellow, Medtronic Neuromodulation

Rainer Goebel, PhD, Professor of Cognitive Science, Maastricht University

Brian Litt, MD, Professor of Neurology, University of Pennsylvania

Helen Mayberg, MD, Professor of Psychiatry, Neurology, and Radiology, and Dorothy C Fuqua Chair of Psychiatry Neuroimaging and Therapeutics, Emory University

Alvaro Pascual-Leone, MD, PhD, Professor of Neurology and Director of the Berenson-Allen Center for Noninvasive Brain Stimulation, Harvard University

Bruce Rosen, MD, PhD, Professor in Radiology, Harvard Medical School, and Director of the Athinoula A Martinos Center for Biomedical Imaging, Massachusetts General Hospital

Nicholas Schiff, MD, Director of the Laboratory of Cognitive Neuromodulation, Cornell University

Stephen Smith, PhD, Associate Director of the Centre for Functional MRI of the Brain and Professor of Biomedical Engineering, University of Oxford

Doris Tsao, PhD, Assistant Professor of Biology and Computation and Neural Systems, California Institute of Technology

David Van Essen, PhD, Professor of Anatomy and Neurobiology, Washington University

www.ingramcontent.com/pod-product-compliance
Lightning Source LLC
Chambersburg PA
CBHW081226170526
45165CB00009B/2968